STRUCTURES FOR ARCHITECTS

Other titles in the series

Fluid Mechanics
Materials and Structures
Soil Mechanics
Structural Steelwork
Surveying

STRUCTURES FOR ARCHITECTS

Second edition

BRYAN J.B. GAULD

B.Sc., M.Sc., M. Phil., DIC, C.Eng., MICE, AIWSc

Senior Lecturer at the School of Architecture, Kingston Polytechnic

Longman Scientific & Technical,
Longman Group UK Limited,
Longman House, Burnt Mill, Harlow,
Essex CM20 2JE, England
and Associated Companies throughout the world.

First published 1984
Second impression 1986
Second edition 1988
Fourth impression 1991

British Library Cataloguing in Publication Data

Gauld, Bryan J.B.
Structures for architects. — 2nd ed.
1. Structures. Theories
I. Title
624.1'7

ISBN 0-582-03727-1

Printed in Malaysia
by Peter Chong Printers Sdn. Bhd.,
Ipoh, Perak Darul Ehsan

CONTENTS

AUTHOR'S PREFACE

This book has been prepared with the object of helping the student to gain the essentials of basic structures. The subject-matter has therefore been restricted and many topics not central to the worked examples have been omitted. For further information, the reader is referred to the wealth of available technical journals and textbooks.

The worked examples have been selected to show the type of structural problems the recently qualified architect meets when the district surveyor or building control officer demands calculations. These examples have been introduced in an effort to reduce the gap between theory and practice.

It is also hoped that this book will help the student to arrive at grand structural schemes, marrying good design and planning with good structural concepts. From the architect's point of view, such concepts will require basic understanding and a knowledge of the 'rules of thumb'. Any calculations for such structures will require the participation and advice of structural engineers.

On the other hand, small projects will require the same basic understanding, the same 'rules of thumb', but the calculations should be well within the capabilities of the architect. These calculations are required to check the safety of the structure, and as they are usually to be checked by a third party who may not be familiar with the project, it is necessary for any calculations to be clearly presented. The reader is therefore recommended to follow the procedure and layout used in the worked examples.

B.J.B.G.

ACKNOWLEDGEMENTS

We are indebted to the following for permission to reproduce copyright material:

W. H. Freeman and Company for our Fig. 1.7 from 'The compound eye of insects' by G. Adrian Horridge. Copyright © 1977 Scientific American, Inc. All rights reserved.

Extracts from British Standards are reproduced by permission of the British Standards Institution, 2 Park Street, London, W1A 2BS from whom complete copies can be obtained.

CHAPTER 1

INTRODUCTION TO BUILDING STRUCTURES

The essence of this book is to establish the sizes of structural elements within a building when the elements are made from timber, steel, masonry and concrete. At the same time the overall concept of stability, form and function are covered to demonstrate how good design is related to good structural decisions.

What is meant by structural decisions and structural design is the process of arriving at a suitable system to support a form or shape and to prevent it from collapsing. The support system is called the structure and the structural elements are those individual parts of the structure which help to support the form.

Before the reader works through the simple calculations found in later chapters, fundamentals of building support systems are explored using examples from Nature. This will help to generate an understanding and feeling for structural principles which can be readily applied to the most complex and the most simple problems, and a realisation that good structural design is related to common sense rather than complex mathematical equations.

1.1 STRUCTURAL REQUIREMENTS OF FORM AND FUNCTION

Much can be learnt from the structural form and function found in Nature. Everywhere we see the phenomena of adaptation producing mechanical fitness to cope with some particular function. Outward forms and inner structures have a perfect comprehension confirming that Nature makes all things with a purpose.

To understand these principles, it is necessary to examine the mechanical properties used by Nature. The materials will have to be strong enough to resist all the natural forces that may be imposed during a lifetime. These forces will take many forms but in simple engineering terms there are two kinds of strengths; strength to resist compression and strength to resist tension. Compression members need to be stiff so that they do not buckle, while tension members require no stiffness but only strength to resist the pull. The structure of a bone is stiff and meets the compression requirement, while the tendon of a muscle is excellent in resisting tension.

FIG. 1.1. Stability of a tree

Running parallel to strength is the requirement for stability. The natural form must have balance and internal stiffness. The roots of a tree prevent the tree from falling over (Fig. 1.1). The roots extend even further than the branches, thus providing a sound base for resisting the wind on a windy day when all the branches are moving to and fro. The tree also has a certain amount of internal stiffness, preventing the trunk and branches being flattened by the wind. In fact, many trees are very cunning as they shed their leaves during the windy months of the year thereby reducing their resistance to the wind.

With animals, their legs provide stability by being placed in such a way as to procure balance during running, walking or standing still. The internal stiffness is achieved by a combination of bones and muscles and to a certain extent, the tautness of the skin. The inner functions have to be protected and not be deformed so that the outward form must be stiff enough to maintain its form and not collapse when exposed to any stress. For example, the rib cage of any animal must be stiff enough to prevent the lungs from being crushed, and the skull must be stiff enough to protect the brain.

Man-made structures have to obey the same fundamental natural laws. They must have enough strength to resist compression and tension and have enough stiffness to be stable both externally and internally. The building must not be blown away, overturned, collapse during a storm, or fall to bits when overcrowded. If it is subjected to earthquake tremors, hopefully it will survive without any loss of life. In short, the building should meet all the basic requirements, and in particular all the basic structural requirements.

To delve more deeply into structural requirements, it is necessary to ask the question, 'what are buildings for?' If this is answered in a very simple way, it could be said that a building encloses a space and protects this space from the natural elements. The fabric which wraps around the building provides the protection. This can be in the form of bricks, blocks, concrete, timber, steel, reinforced plastics, glass, etc. Then, if the materials to be used as the fabric have good structural properties, then three fundamental questions are raised:

(*a*) Should the building fabric be self supporting?

(*b*) Should the fabric be supported by an independent framed structure?

(*c*) Should a combination of (*a*) and (*b*) be used?

The structural requirements for each approach are different, and have to be studied, although the fundamental ones of strength, function and stability are the same.

At this point, it is appropriate to return to examples from Nature. The examples of a rabbit and a tortoise shown in Figs 1.2 and 1.3 are good examples to illustrate the differences between a framed structure and a fabric supporting structure, questions (*a*) and (*b*). The rabbit is the example of a form or shape having an internal framed support system, while the tortoise is the example of a support system generated by the stiffness of the fabric.

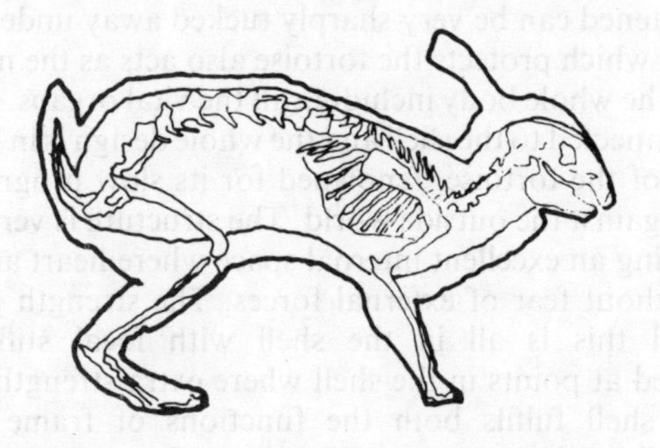

FIG. 1.2. Cross-section of a rabbit (internal frame structure)

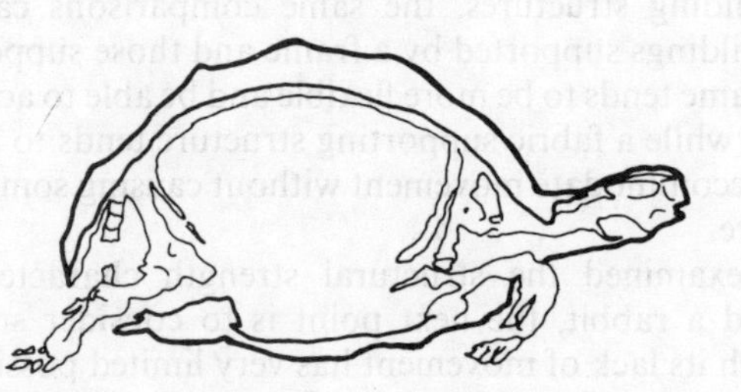

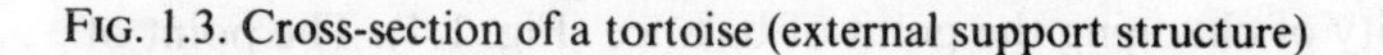

FIG. 1.3. Cross-section of a tortoise (external support structure)

The rabbit has a fur coat or outer fabric to contain all the internal functions and to keep it warm. The vital organs are protected by the rib cage which in turn is supported by the spine which in turn is supported by the legs. The brain, the most sensitive of the organs, is well protected by the skull. The head itself is supported by the neck in the same way as a cantilevered beam supports a load requiring considerable strength where the neck meets the main part of the body. This strength is achieved by a combination of bones slotted together in the form of the spine and muscles which wrap around the bones to keep everything together. The design of the whole skeleton of the rabbit is sensitive to the demand of strength at every point. The skeleton constantly changes in shape and size, only using the amount and size of bones and muscle which are really necessary, but at the same time allowing freedom of movement. It is a computerised design to the very highest standard.

All this structural support keeps the rabbit well clear of the ground and allows it to carry out all its functions of running, hopping, sitting, eating, etc. The tortoise on the other hand is close to the ground and does not have a great deal of movement. He has a shell which protects him from his enemies and a soft head which normally protrudes, but when threatened can be very sharply tucked away under his shell.

The shell which protects the tortoise also acts as the main support system for the whole body including all the vital organs. The legs and neck are connected to the shell and the whole design is in keeping with the nature of the tortoise, renowned for its slow progress, but well protected against the outside world. The structure is very strong and stiff providing an excellent internal space where heart and lungs can operate without fear of external forces. The strength generated to support all this is all in the shell with local stiffeners being incorporated at points in the shell where extra strength is required. Thus, the shell fulfils both the functions of frame and fabric. However, all this is at the cost of freedom of movement, which in the case of the tortoise does not matter, as this lack of movement is one of the parameters of his design.

With building structures, the same comparisons can be made between buildings supported by a frame and those supported by the fabric. A frame tends to be more flexible and be able to accommodate movement, while a fabric supporting structure tends to be rigid and unable to accommodate movement without causing some damage to the structure.

Having examined the structural strength characteristics of a tortoise and a rabbit, the next point is to consider stability. The tortoise with its lack of movement has very limited possibilities. The legs protrude directly out from under the edges of the shell with no possibility of changing the point of balance (Fig. 1.4). The point of

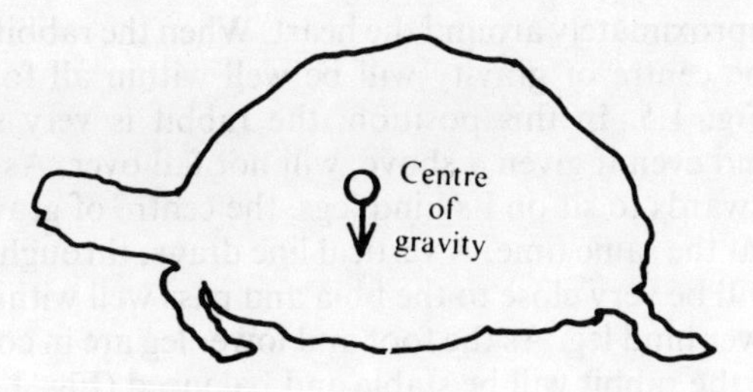

FIG. 1.4. Stability of a tortoise

balance is all to do with the centre of gravity. The centre of gravity is the point within a body where all the weight can be considered to act, no matter what position the body may be in. It is the fulcrum of the point of balance. With a tortoise, the centre of gravity is approximately in the middle of its body, moving slightly forward when the tortoise pushes its head out from under its shell. No matter what the tortoise does with its legs, the centre of gravity stays well within the area of the four feet, and is always stable. If the tortoise is unfortunate enough to topple over on to its back, there is nothing it can do to regain stability.

On the other hand, a rabbit, being more flexible, can easily recover from a position of unbalance such as being on its back. However, a more normal position for the rabbit is to be either standing on all fours or sitting back on its hind legs. It is therefore, worth taking a closer look at the stability of the rabbit in these two positions (Figs. 1.5 and 1.6).

The key to the stability of a rabbit is again connected with the idea of centre of gravity or with the point of balance as discussed in connection with the tortoise. In the case of a rabbit, the centre of

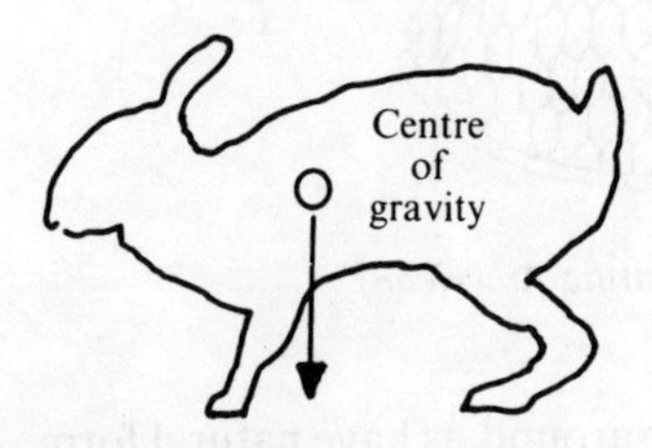

FIG. 1.5. Stability of a rabbit in standing position

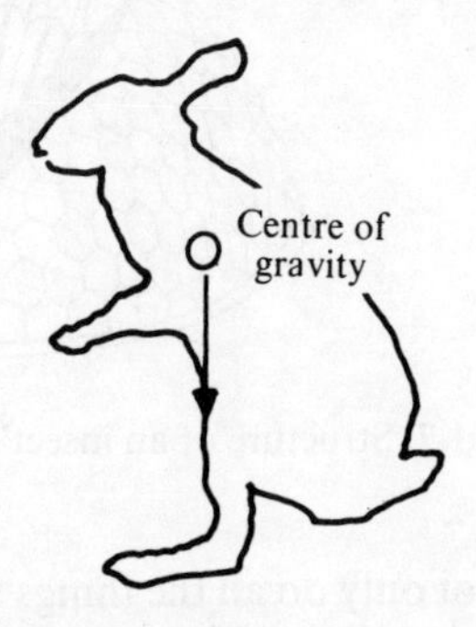

FIG. 1.6. Stability of a rabbit in sitting position

gravity is approximately around the heart. When the rabbit stands on all fours, the centre of gravity will be well within all four feet, as shown in Fig. 1.5. In this position, the rabbit is very stable and balanced, and even if given a shove, will not fall over. As the rabbit moves backwards to sit on its hind legs, the centre of gravity moves backwards at the same time. A vertical line drawn through the centre of gravity will be very close to the fibia and pass well within the hind foot and lower hind leg. As the foot and lower leg are in contact with the ground, the rabbit will be stable and balanced (Fig. 1.6).

The rabbit is stable sitting on its hind legs only because of the very clever design of its hind legs. It is no accident that these legs are long and bend forward into the stomach. In the same way as it is possible to balance a ruler at its centre, the 'knee joint' of the rabbit acts as a fulcrum, a point about which the body of the rabbit is balanced. There is no strain when it sits back on its hind legs as it is balanced and relaxed. By comparison, the tortoise has short straight legs with no possibility of balance other than on all fours.

Moving away from rabbits and tortoises, there is a limitless supply of examples to consider. Every living thing has structural form and as Nature designs with a purpose, the structural design has been finely developed. However, one point of warning at this stage if natural structures are to be compared to man-made structures. Nature uses live materials while man uses dead materials, and the two do not always behave in the same way. For example, the soles of shoes wear out, but the soles of bare feet grow thicker with excessive use.

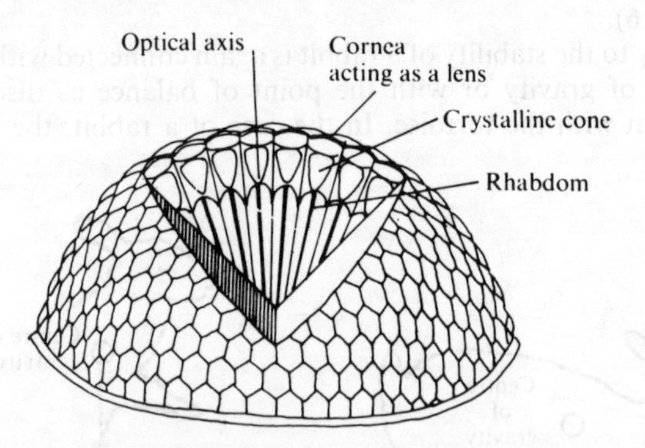

FIG. 1.7. Structure of an insect's eye (*Scientific American*)

Not only do all the things we can see around us have natural form, but also all the things we can hardly see. The structure shown in Fig. 1.7 is the structure of an insect's eye magnified many times. An insect has many corneas in each eye which act as lenses. Light travelling through the cornea is focused through a transparent cone on to a light-sensitive element called the rhabdom. All these elements have to be held together and be supported. This is all done by a geodesic dome grid on the outside of the eye as illustrated in Fig. 1.7, which by its very nature takes up the shape of a hemisphere.

Each node or joint in the grid is held in position by three 'rods', and as these 'rods' are relatively short, they are strong both in tension and compression. Thus the whole structural framework of the domed grid is held firmly in position, giving the corneas a stiff and strong support, enabling the insect's eye to function efficiently.

The well-known mathematician, educationalist and engineer, Buckminster Fuller, has used this simple framework in many of his famous man-made structures, and refers to them as geodesic domes. His concepts are precisely the same as the structure of an insect's eye. This illustrates how many architects and engineers have copied basic natural forms to create magnificent structures and buildings, many of which have become famous. For the imaginative designer there are still many concepts used by Nature which have not yet been fully explored by man.

It is quite possible to discuss many other structural forms exploited by Nature, but as has been stated before, that is not the intention of this book. Hopefully, the deviation into the wonders of Nature has encouraged the reader to use his eyes and common sense to observe the natural world and apply this experience to general structural decisions, even if those decisions are quite small.

1.2 STRENGTH, STABILITY AND SERVICEABILITY

In the previous section, it has been emphasised that the structure of a tortoise and a rabbit must have sufficent strength and stability to behave in a way as to carry out its everyday functions. Engineers call this last requirement 'serviceability'. Why this word was chosen rather than the word 'function' is not too clear, as the term 'serviceability' is defined as being able or willing to serve. To be fair, the structure does behave as a faithful servant and does render long service to the occupier. However, if the reader does not like the term 'serviceability', then the term 'function' can be substituted.

Therefore, as in the case of natural forms, building structures require to have strength, stability and serviceability. It is the responsibility of the designer to meet all three requirements. Each has to be analysed separately and each one must meet the minimum standards, so that:

1. *Strength*: The structural members must be strong enough to

resist all the forces imposed on the building during its lifetime. This may mean that the forces involved may pull a structural member under one condition and push it under another, or even twist it under yet another. The designer has to consider the worst possible combination of forces or 'loads', and analyse how this extreme load will affect the structural strength of the member.

2. *Stability*: The building must not topple over, blow away or collapse due to lack of stiffness. The building must have sufficient bracing so that the walls remain at right angles to the floor.

3. *Serviceability or function*: The occupants must find the building comfortable and usable. The structural members must be stiff, so that the floors remain horizontal and the building does not sway too much. Furniture must not slide into the middle of the room due to excessive deflection of the floor, and in a tower block, the water in a basin must not be thrown from one side to the other by the sway of the building. The building must feel and look safe to the layman.

A building may well meet one or two of the requirements, but unless it meets all three, it is a failure. For example, a beam might have sufficient strength, but if it has excessive deflection, it will fail under serviceability. So it must be emphasised again that the building must meet all three requirements.

1.3 PHILOSOPHIES OF STRUCTURAL DESIGN

It seems at this point appropriate to discuss a number of approaches to structural design.

The growth design method

The unique approach Nature has at arriving at structural form and function has been demonstrated in Section 1.1. The structure seems to grow into the right shape with the appropriate strength as the growth pattern of Nature is able to adapt to the surrounding natural forces. For example, if a load is placed on the branch of a tree, that branch will slowly adapt and grow stronger to cope with the extra load. In comparison, man's efforts, although commendable, do not compare with this kind of skill.

Experience, intuition, common sense, and rules of thumb

The experienced carpenter, bricklayer or builder has a wealth of experience built up over the years from working on many buildings. If intuition and common sense are added, these people are able to size most structural members in a building without having to resort to design codes and mathematics. They have an understanding of what the correct size should be. For the less experienced, it is not possible to have a sense of structural sizes, so they have to rely on rules of thumb. These rules are fully discussed in the next section (1.4) and it is sufficient at this stage only to appreciate that they can be used to obtain the approximate sizes of structural members. The rules are also useful to engineers as a first approximation to a problem before a computerised analysis is carried out. The rules, intuition and common sense are part of the initial structural judgement which produces a good concept and a good design. The computers and codes are only there to confirm this judgement.

Mathematics: Elastic method

Good structural materials behave elastically. That is, when they are pulled, pushed, bent or twisted, they do not suffer any permanent damage. When the forces causing these disfigurements are removed the structure returns to its original shape. If the forces on the material are increased to the point at which the material begins to suffer permanent distortion then the material is no longer behaving elastically. The point at which the permanent deformation occurs is called the yield point. This is the point at which the material begins to yield. This is discussed in more detail in section 2.4, and it needs only be appreciated at this stage that the elastic approach to structural design first finds out what the yield point of the material is and then says the permissible safety for the material will be the yield strength divided by a safety factor. The strength of each and every structural member in a building must never exceed this permissible safety.

So for elastic design:

$$\text{Permissible safety} = \frac{\text{Yield point of material}}{\text{Safety factor}}$$

Mathematics: Limit state design

The ultimate strength of a material is the point at which it breaks. The ultimate strength of a building is the strength of the structure just before the building falls down. Although the building will look very sorry for itself long before this with sagging beams and floors, disaster will not strike until the building collapses. There is a latent strength between the yield point as discussed in the explanation of elastic design and the ultimate strength or point of collapse. The limit state approach to structural design argues, why not use this ultimate strength when considering the overall safety of the building and apply the safety factors to the ultimate strength rather than the yield point conditions.

There are two variables to be considered, the variation of the loads applied to the structure and the variations in the strength of the materials used. It is, therefore, necessary to apply safety factors to both the loads and the materials so as to achieve a structural design of acceptable probabilities regarding safety and function.

So for limit state design:

$$\text{Permissible safety} = \left(\frac{\text{Ultimate strength of materials}}{\text{Safety factor}}\right)$$
$$+ (\text{Applied loads} \times \text{Safety factor}$$

The elastic approach is used in Chapters 6, 7 and 8 for timber, steel and masonry, while the limit state approach is illustrated in Chapter 9 covering reinforced concrete. Both these methods use safety factors, and it should be remembered that there is a certain amount of intelligent guesswork as to establishing the safety of structures. Safety factors are arrived at by a number of judgements. British Standard Code of Practice CP 110 confirms this by saying 'values are based on statistical evidence or on an appraisal of experience'. In other words guesswork, but guesswork with an intimate appreciation of the problems.

1.4 RULES OF THUMB, APPROXIMATE SIZES OF STRUCTURAL MEMBERS

The most important structural information an architect requires during the planning stage of a building project is a conceptual idea and understanding on how the building is to be structured. At the detailed planning stage, the most important structural information is the approximate sizes of the structural members. If this information is available, then decisions such as headroom, etc., can be made without disruption to the planning and design at a later date. The experienced architect and engineer has an instinctive feeling for structural form gained from experience over a number of years, but the student has to rely on a number of rules of thumb. Tables 1.1 and 1.2 set out a number of rules to cover beams, portal frames, cantilevers, floor slabs and columns.

The approximate depth of structural members is mainly a function of serviceability, rather than strength of stability. A beam can be made stronger by increasing the amount of material and can be made stable by providing sufficient lateral bracing. However, to make a building serviceable, the beam must not be too flexible, otherwise ceilings will crack and furniture will slide on the sloping floor into the middle of the room. Therefore, deflections must be limited and the most efficient way to do this is to make sure that the beam has sufficient depth. The span to depth ratios given in Table 1.1 provide a surprisingly good 'guesstimation' as to the most efficient depths of beams and slabs regardless of the material used.

It must be emphasised that these are rules of thumb, and what these span to depth ratios give in general are the most economic sections regarding the weight of the beam to the span. However, there are often design parameters which may change this. For example, if a steel beam is to support a timber floor, it may be prudent to lose the beam within the depth of the timber joists by using a heavy Universal Column section as a beam, and thus achieve a flush ceiling, as shown in Fig. 1.8. The weight of the steel beam may be as much as four times greater than the economic section but the shallower section allows the architect to avoid having downstand beams. It can also be argued that the shallower section may save one or two courses of brickwork and hence produce a more economical solution as a whole.

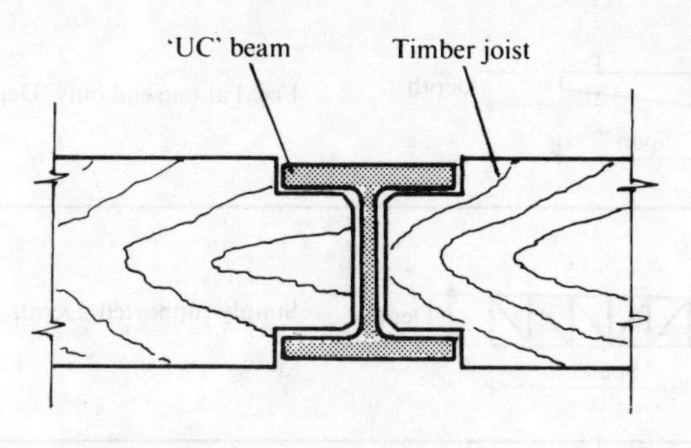

FIG. 1.8. Universal column used as a shallow beam

The rules of thumb as set out in Tables 1.1 and 1.2 are illustrated in Worked Examples 1.1–1.5. Some judgement has to be made as to whether the loading on the beam is heavy, medium or light. Otherwise it is simply a matter of applying the recommended span to depth ratios.

FURTHER READING

Thompson, D'Arcy (1917) *On Growth and Form*, volumes I and II. Cambridge University Press.
Hogden, Lancelot (1936) *Mathematics For the Million*. George Allen and Unwin, London.

TABLE 1.1 Approximate sizes of structural members

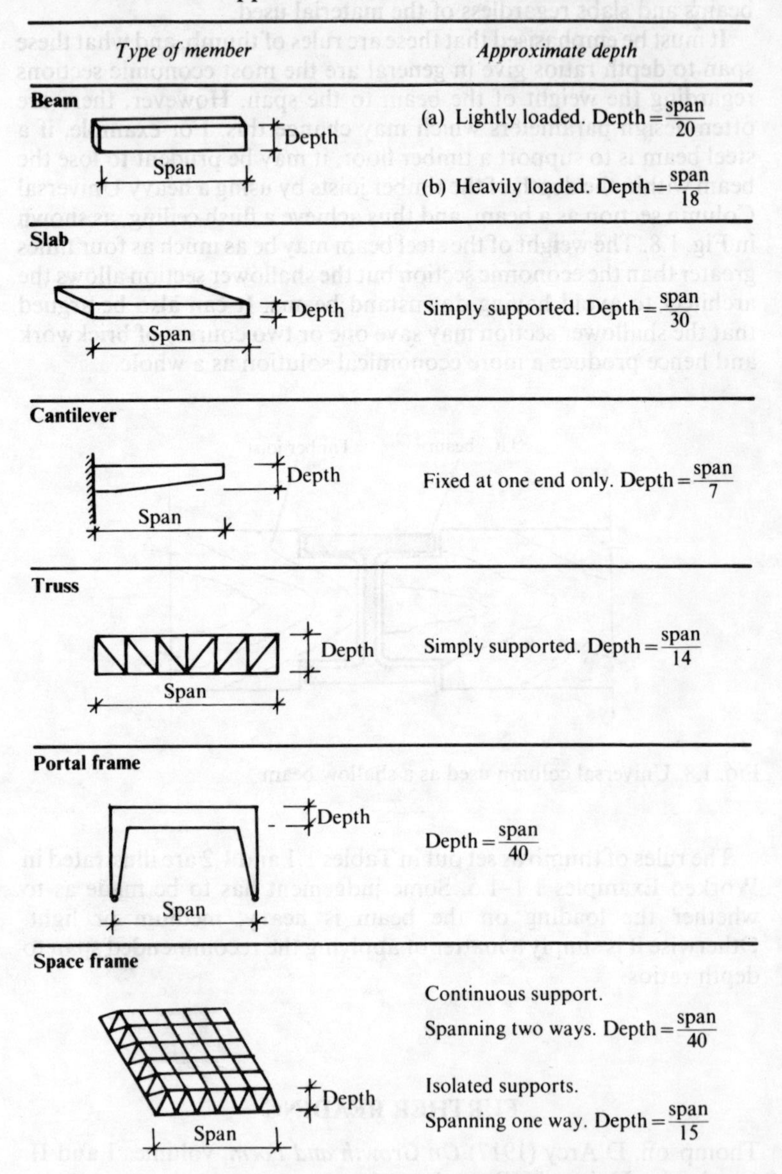

Type of member	*Approximate depth*
Beam	(a) Lightly loaded. Depth $= \frac{\text{span}}{20}$ (b) Heavily loaded. Depth $= \frac{\text{span}}{18}$
Slab	Simply supported. Depth $= \frac{\text{span}}{30}$
Cantilever	Fixed at one end only. Depth $= \frac{\text{span}}{7}$
Truss	Simply supported. Depth $= \frac{\text{span}}{14}$
Portal frame	Depth $= \frac{\text{span}}{40}$
Space frame	Continuous support. Spanning two ways. Depth $= \frac{\text{span}}{40}$ Isolated supports. Spanning one way. Depth $= \frac{\text{span}}{15}$

TABLE 1.2 Approximate sizes of columns

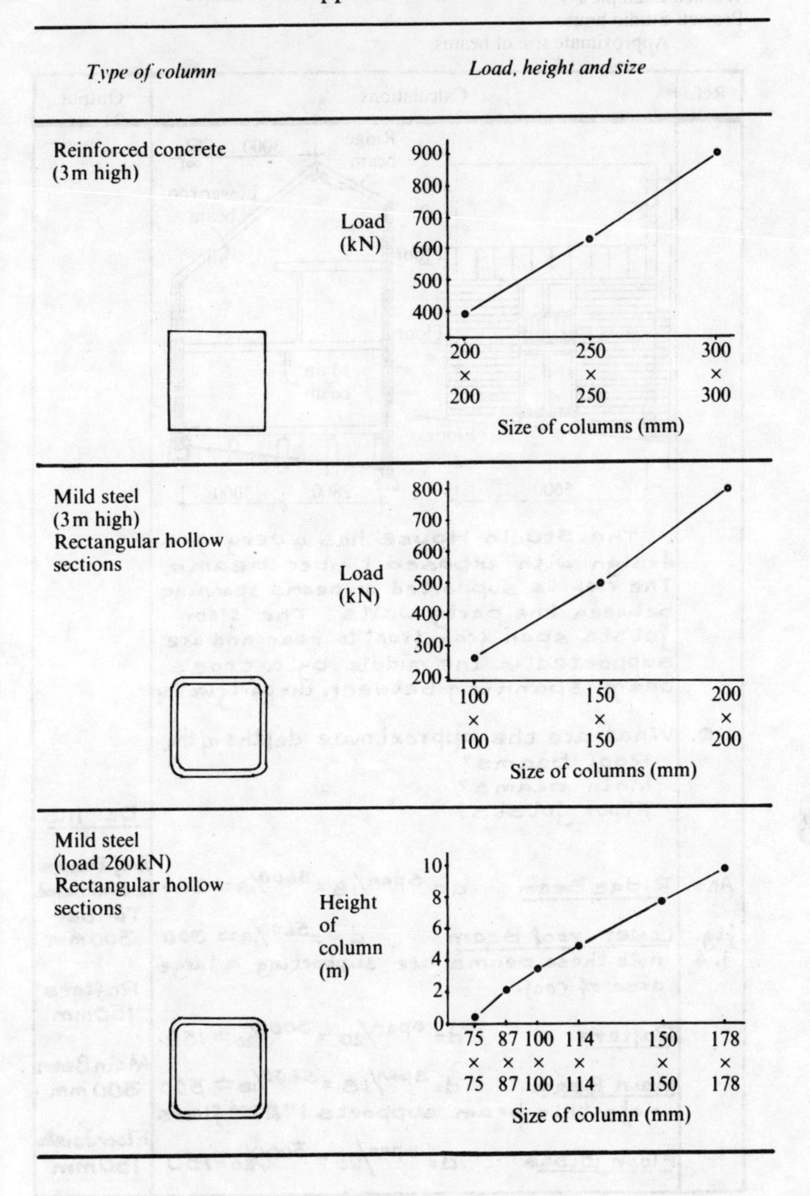

Worked Example 1.1
Project: Studio house
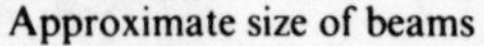
Approximate size of beams

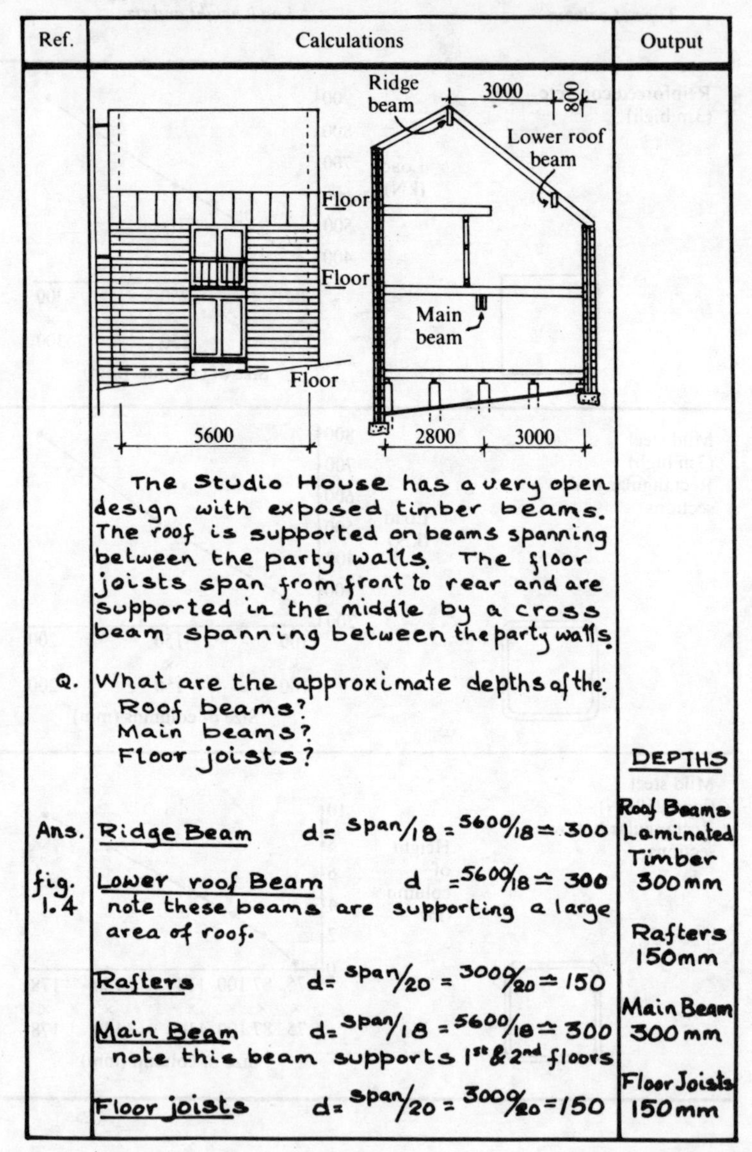

Ref.	Calculations	Output
	The Studio House has a very open design with exposed timber beams. The roof is supported on beams spanning between the party walls. The floor joists span from front to rear and are supported in the middle by a cross beam spanning between the party walls.	
Q.	What are the approximate depths of the: Roof beams? Main beams? Floor joists?	DEPTHS
Ans.	Ridge Beam $d = \text{span}/18 = 5600/18 \doteq 300$	Roof Beams Laminated Timber 300 mm
fig. 1.4	Lower roof Beam $d = 5600/18 \doteq 300$ note these beams are supporting a large area of roof.	
	Rafters $d = \text{span}/20 = 3000/20 \doteq 150$	Rafters 150 mm
	Main Beam $d = \text{span}/18 = 5600/18 \doteq 300$ note this beam supports 1st & 2nd floors	Main Beam 300 mm
	Floor joists $d = \text{span}/20 = 3000/20 = 150$	Floor Joists 150 mm

Worked Example 1.2
Project: Terrace house
Approximate sizes of joists and beams

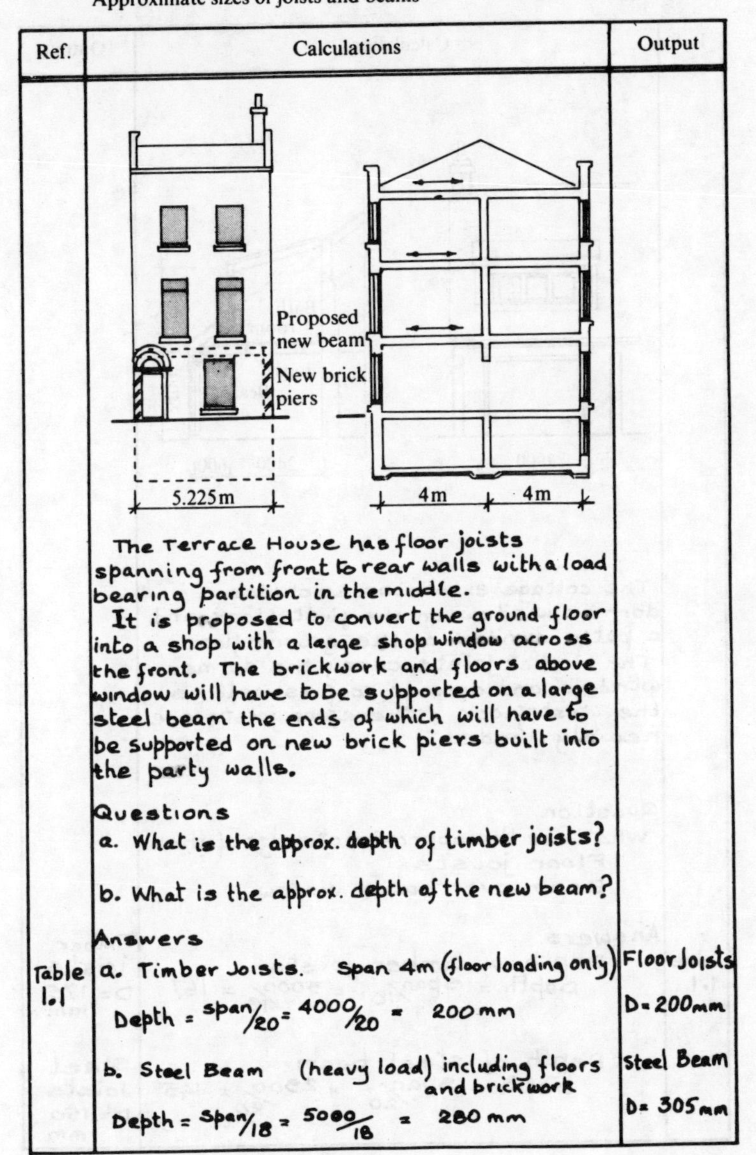

Ref.	Calculations	Output
	The Terrace House has floor joists spanning from front to rear walls with a load bearing partition in the middle. It is proposed to convert the ground floor into a shop with a large shop window across the front. The brickwork and floors above window will have to be supported on a large steel beam the ends of which will have to be supported on new brick piers built into the party walls.	
	Questions a. What is the approx. depth of timber joists? b. What is the approx. depth of the new beam?	
	Answers	
Table 1.1	a. Timber Joists. Span 4m (floor loading only) Depth = span/20 = 4000/20 = 200 mm	Floor Joists D = 200mm
	b. Steel Beam (heavy load) including floors and brickwork Depth = Span/18 = 5000/18 = 280 mm	Steel Beam D = 305mm

Worked Example 1.3
Project: Cottage extension
Approximate sizes of beam and joists

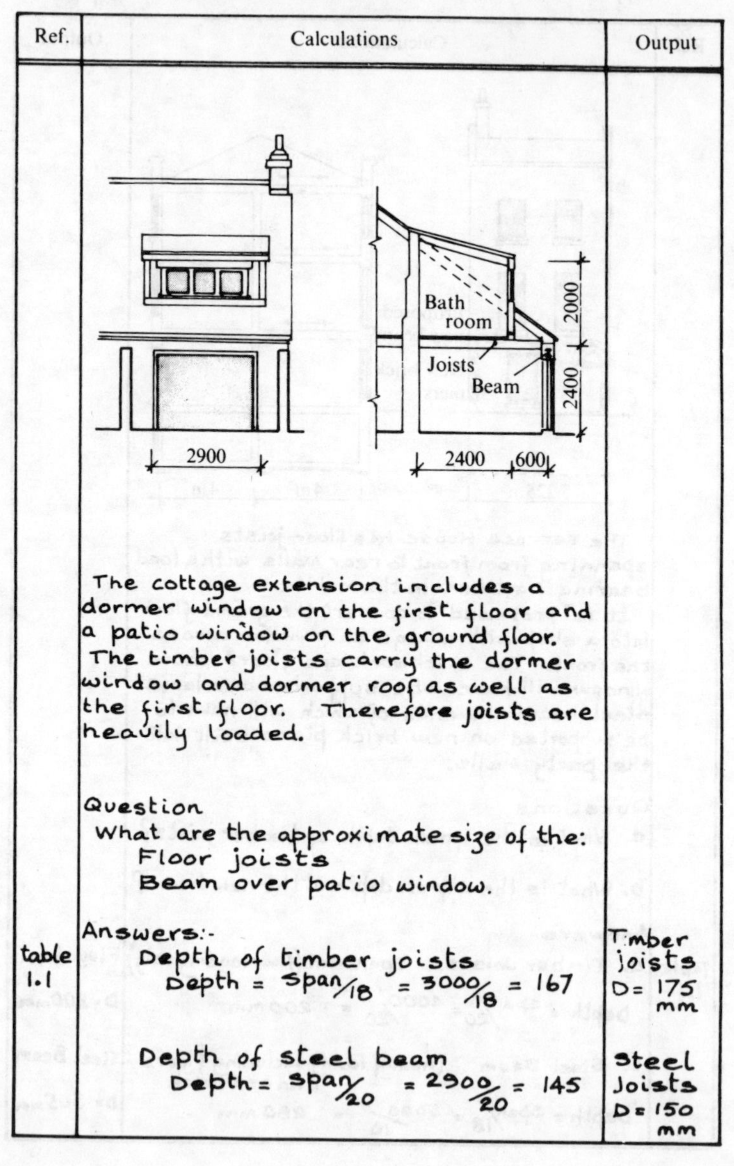

Ref.	Calculations	Output
	The cottage extension includes a dormer window on the first floor and a patio window on the ground floor. The timber joists carry the dormer window and dormer roof as well as the first floor. Therefore joists are heavily loaded.	
	Question What are the approximate size of the: Floor joists Beam over patio window.	
table 1.1	Answers:- Depth of timber joists $\text{Depth} = \text{Span}/18 = 3000/18 = 167$	Timber joists D = 175 mm
	Depth of steel beam $\text{Depth} = \text{Span}/20 = 2900/20 = 145$	Steel Joists D = 150 mm

Worked Example 1.4
Project: Shop extension
Approximate sizes of beams

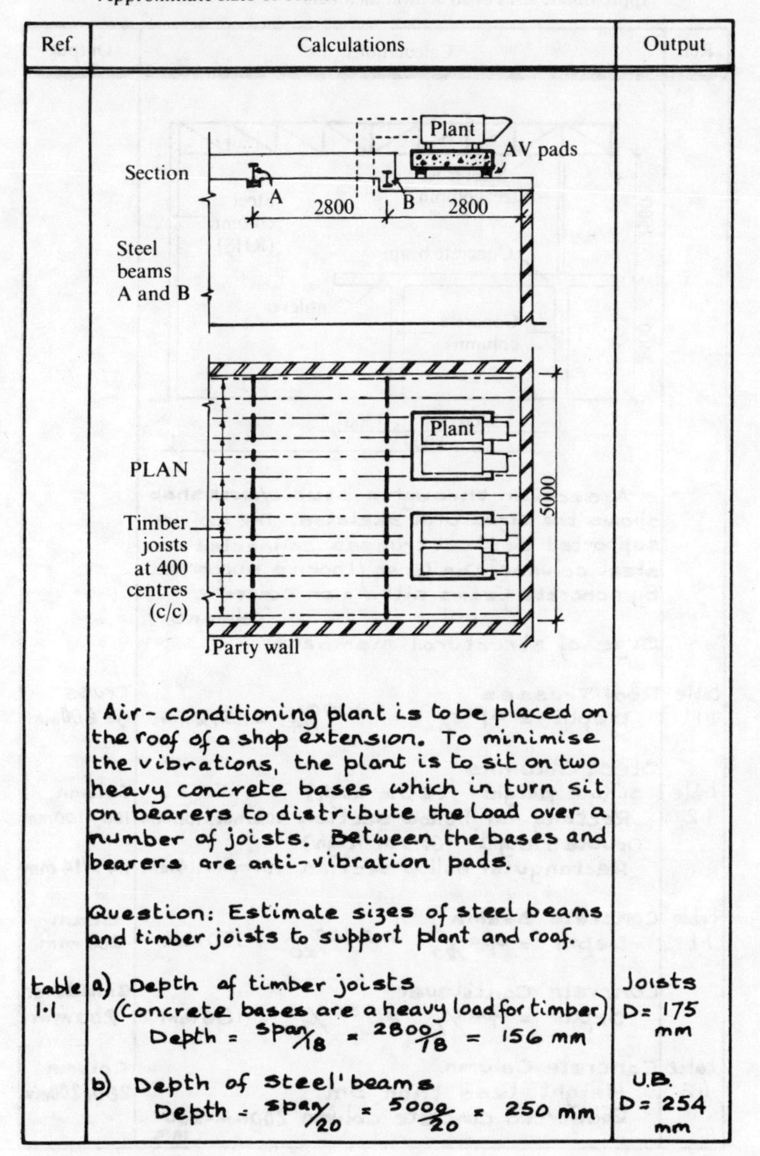

Ref.	Calculations	Output
	Air-conditioning plant is to be placed on the roof of a shop extension. To minimise the vibrations, the plant is to sit on two heavy concrete bases which in turn sit on bearers to distribute the load over a number of joists. Between the bases and bearers are anti-vibration pads.	
	Question: Estimate sizes of steel beams and timber joists to support plant and roof.	
table 1·1	a) Depth of timber joists (Concrete bases are a heavy load for timber) Depth = $\text{span}/18 = 2800/18 = 156$ mm	Joists D = 175 mm
	b) Depth of steel beams Depth = $\text{span}/20 = 5000/20 = 250$ mm	U.B. D = 254 mm

Worked Example 1.5
Project: Studio/workshop
Approximate sizes of structural members

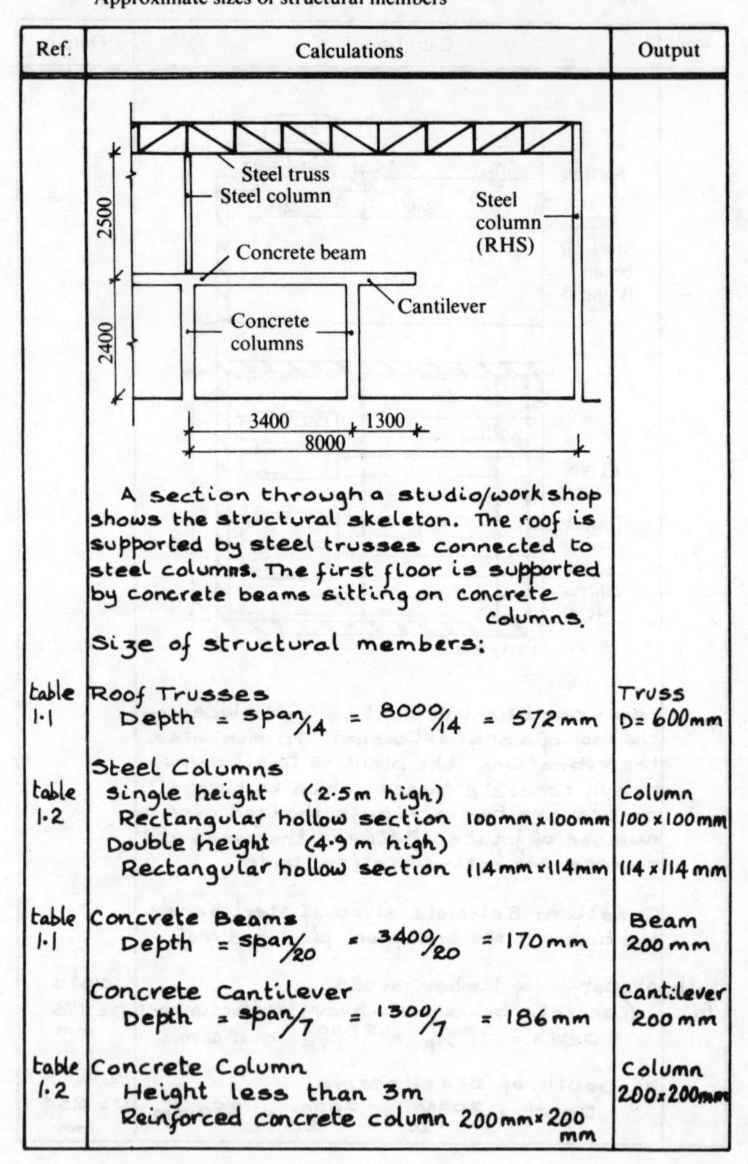

Ref.	Calculations	Output
	A section through a studio/workshop shows the structural skeleton. The roof is supported by steel trusses connected to steel columns. The first floor is supported by concrete beams sitting on concrete columns.	
	Size of structural members:	
table 1.1	Roof Trusses Depth = span/14 = 8000/14 = 572 mm	Truss D = 600 mm
table 1.2	Steel Columns Single height (2·5 m high) Rectangular hollow section 100 mm × 100 mm Double height (4·9 m high) Rectangular hollow section 114 mm × 114 mm	Column 100 × 100 mm 114 × 114 mm
table 1.1	Concrete Beams Depth = span/20 = 3400/20 = 170 mm	Beam 200 mm
	Concrete Cantilever Depth = span/7 = 1300/7 = 186 mm	Cantilever 200 mm
table 1.2	Concrete Column Height less than 3 m Reinforced concrete column 200 mm × 200 mm	Column 200 × 200 mm

CHAPTER 2

BASIC STRUCTURAL CONCEPTS

2.1 FORCES

The unit of force is the newton (N), named after Isaac Newton who observed apples falling from trees, and realised that the force produced by a falling apple is equal to the mass times the earth's gravitational pull. If the apple was on the moon, it would be the moon's gravitational pull. Although the mass would be the same on the earth as on the moon, the force would be different because the gravitational pull of the earth and moon are not the same.

The definition of the newton is the force which when applied to a body having a mass of 1 kilogram, causes an acceleration of 1 metre per second in the direction of application of the force. The earth s gravitational pull is $9.81\,m/s^2$, so 1 kilogram of mass will produce a force of 9.81 N, and for calculations of architectural structures as covered by this book, we can assume that:

$$1\,kg \text{ weight produces a force} = 10\,N$$

2.2 MOMENTS

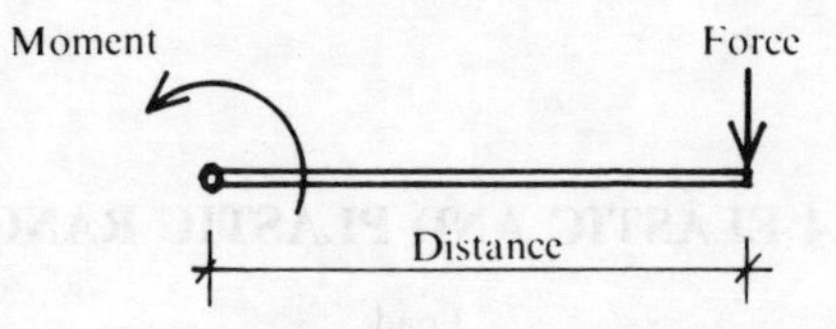

FIG. 2.1

A moment is a force times a distance (Fig. 2.1). The unit of a moment is kilonewton metres, or newton millimetres. It is sometimes necessary to change kN·m into N·mm. The kilonewtons have to be multiplied by a 1000 to bring them to newtons and the metres have to be multiplied by a 1000 to bring them to millimetres. Therefore:

$$1\ kN{\cdot}m = 1\,000\,000\,N{\cdot}mm$$

2.3 STRESS AND STRAIN

A good way to understand the structural meaning of stress and strain is to consider the meaning of these words in everyday life. For example, consider a group of people in a design office. If they are under stress, it may mean that a project has to be finished by a certain date. The deadline will be an external force, exerting a stress on the group. The intensity of the stress will depend upon the size of the group. The smaller the group, the more work each individual will have to do, causing a greater intensity of stress. As a result, the individual may suffer a physiological change, and will be under strain. The strain is internal while the stress is due to the external force.

The same definition can be applied to the words stress and strain as used in structures. The stress is the intensity of the external force and the strain is the deformation or the change in shape of the material within the structural member.

$$\text{Stress} = \frac{\text{Force}}{\text{Cross-sectional area}}$$

$$\text{Strain} = \frac{\text{Extension}}{\text{Original length}} \quad \text{or} \quad \frac{\text{Change in shape}}{\text{Original shape}}$$

Units:

Stress = N/mm^2

Strain is dimensionless

A tension stress is due to a pulling force and a compression stress is due to a compression or crushing force.

2.4 ELASTIC AND PLASTIC RANGE

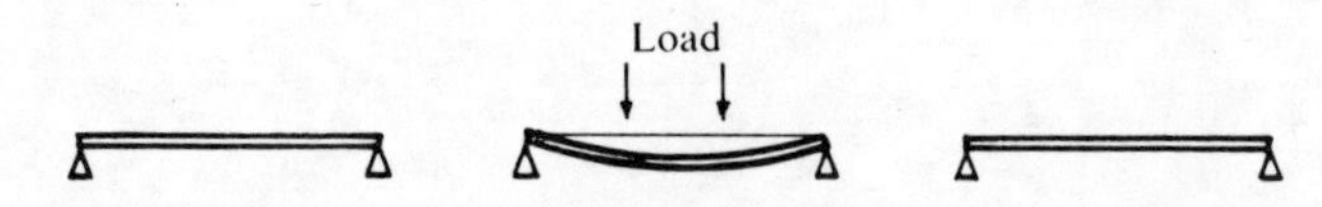

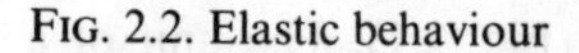

FIG. 2.2. Elastic behaviour

For a material to be suitable for structural use, it must behave in an elastic way. The word elastic is used to describe the recovery potential of the material. This can be illustrated by considering a beam (Fig. 2.2). When a load is applied, the beam will deflect, but if the beam is behaving elastically, it will return to its original position when the load is removed, no matter how many times the load is applied. It is important during the life of a building, say 50 years, for all the structural members to behave in an elastic way with no permanent deformations occurring under the original design loads.

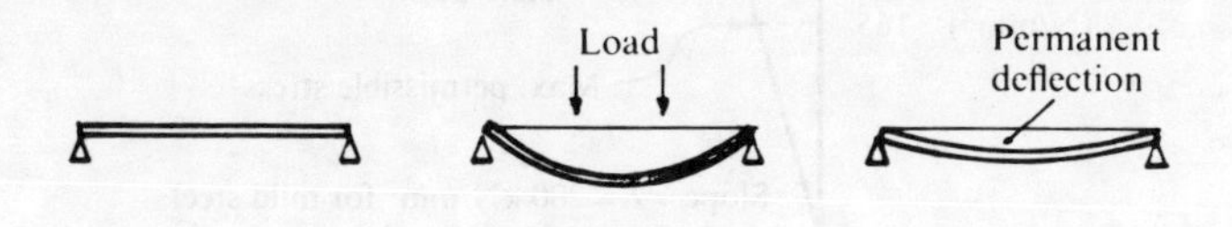

FIG. 2.3. Plastic behaviour

A material is not suitable for structural use if it behaves in a plastic way. For example, when a beam does not return to its original shape after a load has been applied and removed, then plastic deformation has occurred (Fig. 2.3). Every time the load is applied the plastic deformation will increase until a point is reached when the beam will fail. The point of breaking is called the ultimate strength of the material.

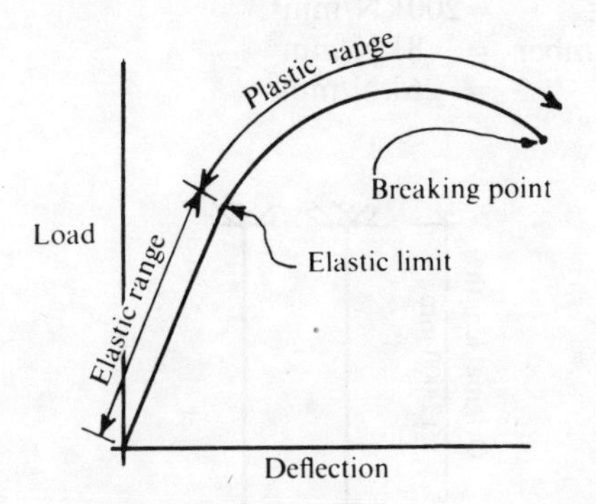

FIG. 2.4. Relationship of load and deflection

Elastic materials usually only behave elastically up to a certain load, after which the material behaves plastically (Fig. 2.4). In a building, it is important that the load is kept within the elastic range as permanent deformations must not occur, but when considering the structural safety of the building, the plastic range may be taken into consideration. The material will not fail until the breaking point is reached although the structure may have deformed well before this point, giving ample warning of failure.

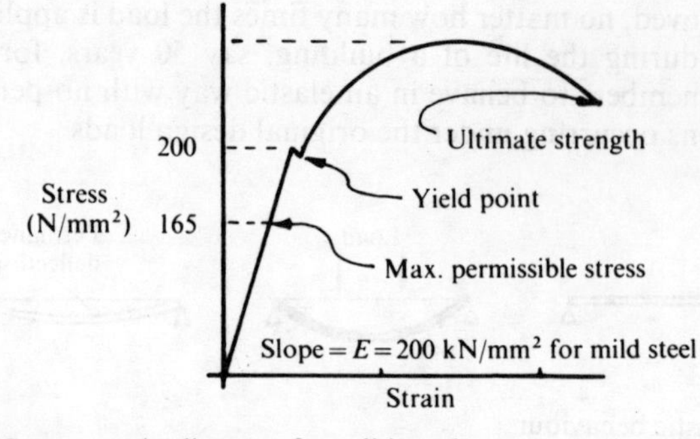

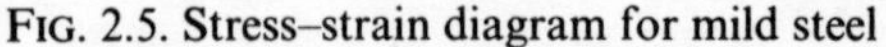

FIG. 2.5. Stress–strain diagram for mild steel

The relationship of stress and strain for mild steel is shown in Fig. 2.5. Up to the yield point, the stress is directly proportional to the strain, beyond this point it is not.

Over the elastic range:

$$\text{Stress varies as strain}$$
$$\text{Stress} = \text{Constant} \times \text{strain}$$
$$= E \times \text{strain}$$

where E = Young's modulus.

For mild steel E = 200 kN/mm²
softwood timber = 8 kN/mm²
concrete = 26 kN/mm²

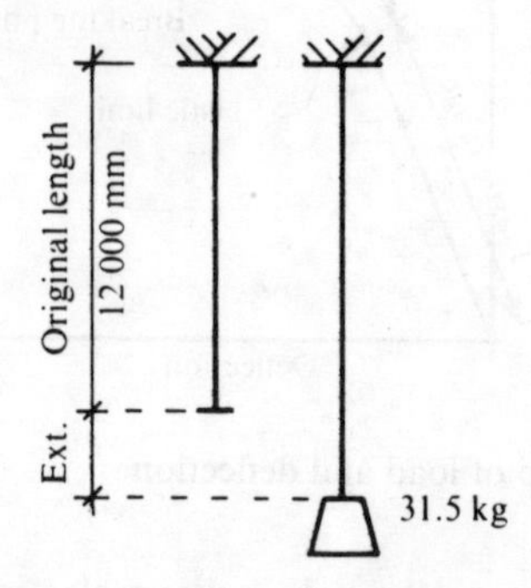

FIG. 2.6

The application of the relationship between stress, strain and Young's modulus, can best be illustrated by an example, such as a piece of wire hanging from a stairwell with a weight on the end. The weight will cause the wire to stretch (Fig. 2.6). If the length and diameter of the wire and the magnitude of the weight are known, then it is possible to predict the amount the wire will extend by using the stress–strain relationship.

If length of wire = 12 m
diameter = 2 mm
weight = 31.5 kg
Assume 1 kg = 10 N
Force = 31.5 kg × 10 = 315 N
Area of wire = $\pi r^2 = 3.14 \times (1\ \text{mm})^2 = 3.14\ \text{mm}^2$

$$\text{Stress} = \frac{\text{Force}}{\text{Area}} = \frac{315\ \text{N}}{3.14\ \text{mm}^2} = 100\ \text{N/mm}^2$$

$$\text{Strain} = \frac{\text{Extension}}{\text{Original length}} = \frac{\text{Extension}}{12000\ \text{mm}^2}$$

Young's modulus for mild steel = $200\ \text{kN/mm}^2$

STRESS = $E \times$ STRAIN

$$100\ \text{N/mm}^2 = 200000\ \text{N/mm}^2 \times \frac{\text{Extension}}{12000\ \text{mm}}$$

Therefore:

EXTENSION = 6 mm

2.5 PRIMARY LOADS

The primary loads which a building has to be strong enough and stiff enough to resist are:

1. Dead load
2. Imposed or live load
3. Snow load
4. Wind load

There are other more specialised loads which a building may be subjected to, but these are dealt with in the next section (2.6), and are described as secondary loads.

The primary loads are the natural forces due to gravity and the dynamic force due to wind. The magnitude of snow and wind loads will depend upon the geographical position of the building and its exposure to the elements. Unlike the dead load which can be calculated quite accurately, estimation of the other three loads involves a certain amount of guesswork as their precise values cannot be calculated. This guesswork is based on 'scientific evidence', so that the designer does not have to do the guessing but simply refer to the appropriate technical reference documents.

The scientific guesses regarding wind loads are quite complex and are beyond the scope of the worked examples in this book, except for good common-sense detailing such as strapping roofs down into the walls and thus preventing the wind from lifting the roof off. Most of the examples cover conversion work, which is the name given to converting old buildings into new or modernised ones. This generally involves using heavy materials which are not affected by the wind, but involves mainly dead and imposed loads only. However, the designer should be well aware of the effects of each load, primary or secondary, and know when to seek the expert advice of a structural engineer, in particular with regard to secondary loads.

Dead load

The weight of the permanent elements of a building such as beams, floor slabs, columns and walls, produce a dead load on the structure. These elements are always there and will always produce the same constant 'dead' load during the life of the building. The volume of these elements can be calculated and if multiplied by the density of the material, the actual dead weight of each element can be calculated. Thus an accurate estimate of the dead load for the whole building can be made.

Dead load then is the weight of all the permanent parts of the building which will include both the structure and the fabric, in fact anything which will not move. The dead load will only change if major alterations are carried out. If this happens, then the design of the whole structure will have to be reconsidered.

The density of various building materials are given in Table 2 in Appendix 3, and estimated dead loads of typical roof, floor and wall sections are given in Table 1 in the same appendix.

Imposed or live load

All the movable objects in a building such as people, desks, cupboards and filing cabinets, produce an imposed load on the structure. This loading may come and go with the result that its intensity will vary considerably. At one moment a room may be empty, yet at another packed with people. Imagine the 'extra' live load at a swinging lively party!

This raises a number of questions such as how many people can squeeze into one room, what happens if they all jump up and down, how often would this occur, and how much load is all this putting on the floor?

With these questions in mind, and a knowledge of the proposed use of the building, an estimation of the imposed load can be made with the help of statistical information. What the building is to be used for is fundamental to this value. For example, if the room is to be used as a stack room in a library, then the imposed load will have to be equated to the weight of a stack of books right up to the ceiling. If a room is to be used as an office, then the imposed load has to be related to the movable objects which would normally be associated with an office. An office should not be used as a stack room, and if it is, the structure should be strengthened to support the extra load. The classic situation is when an office, the structure of which has been designed for normal office usage, is taken over by a Government department. Before long, documents are piled up almost to the ceiling, causing the floors to deflect to such an extent that the doors on the floor below will not open or close!

Imposed loads are not only important for new buildings, but also for existing buildings where there is to be a change of use. Appendix 2 sets out a number of recommended values of imposed loads for a variety of buildings, and if a more comprehensive selection are required, then reference 2.2 should be consulted.

Snow load

The magnitude of the snow load will depend upon the latitude and altitude of the site. In the lower latitudes no snow would be expected while in the high latitudes, snow could last for six months or more. In such locations, buildings have to be designed to withstand the appropriate amount of snow. The shape of the roof also plays an important part in the magnitude of the snow load. The steeper the pitch, the smaller the load. The snow falling on a flat roof will continue to build up and the load will continue to increase, but on a pitched roof a point is reached when the snow will slide off. The steeper the pitch, the sooner this point is reached. Appendix 2, Table 2, gives the recommended values of imposed roof loads to be taken in the British Isles, and they are the same regardless of location. In countries where the climate varies considerably from one end to the other, a system of zones are introduced so that the intensity of the imposed snow loading varies from one zone to the next.

Wind load

Wind has become a very important load in recent years due to the extensive use of lighter materials and more efficient building techniques. A Victorian building built with heavy masonry, timbers and slates will not be affected by the wind load, but the structural design of a modern steel clad industrial building is dominated by the wind load, affecting its strength, stability and serviceability. The wind acts both on the main structure and on the individual cladding units. The structure has to be braced to resist the horizontal load, and anchored to the ground to prevent the whole building from being blown away, if the dead weight of the building is not sufficient to hold it down. The cladding has to be securely fixed to prevent the wind from ripping it away from the structure.

Therefore, careful consideration has to be given to all aspects of the wind load in the design of modern buildings, but the main aspect when considering the building as a whole is how to make it stiff enough to withstand the lateral wind load. Figure 2.7 explains four ways of achieving this. For example, most office blocks are made stiff by using an *in situ* concrete lift shaft as a stiff central core to which all the floors are rigidly fixed, (iii). The floors act as horizontal diaphragms transferring the stiffness from the cores to the whole building. Other methods are to use shear walls (iv), external bracing (i) and rigid joints between beams and columns (ii). For very tall buildings, the rigid joint method can be developed to form a rigid external 'tube'. The external structure instead of having cross bracing, is stiffened by very deep spandle beams and wide columns. This kind of structure is called 'pierced' tube construction, because the building looks like a tube perforated by small holes, the holes being the windows.

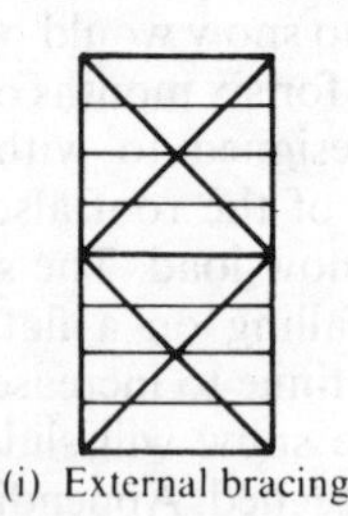

(i) External bracing

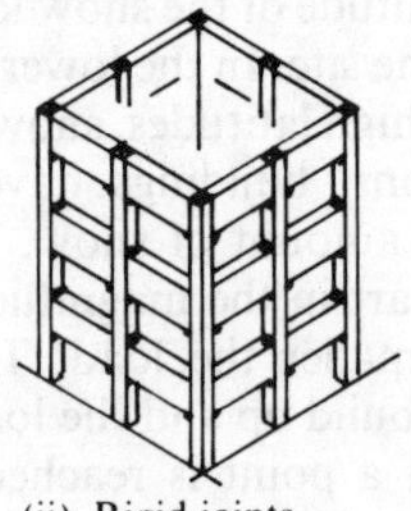

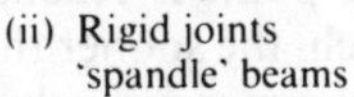

(ii) Rigid joints 'spandle' beams

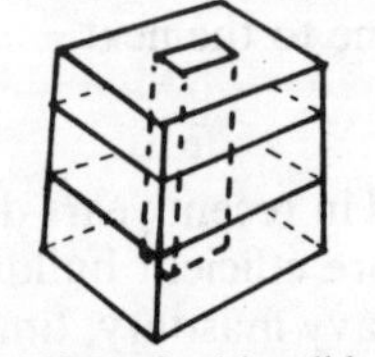

(iii) Lift shaft with solid walls

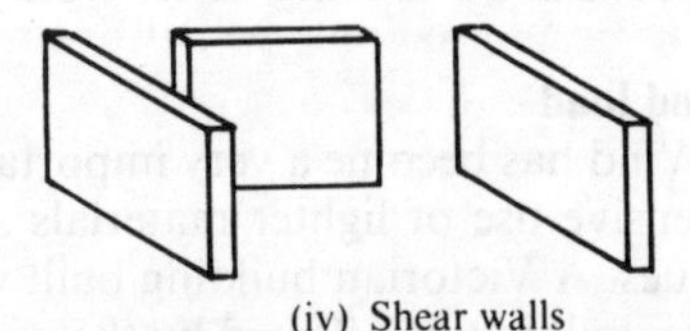

(iv) Shear walls

FIG. 2.7. Bracing a building against a horizontal wind load

Returning to the question of cladding, it should be appreciated that when wind flows around a building, it can produce some very high suction pressures. These occur mainly at the leading edges as illustrated in Fig. 2.8. In these areas, the cladding has to be firmly fixed to the structure and the roof has to be firmly held down. The flatter the roof, the higher the suction forces are, and the more important it is to make sure that the holding-down straps are securely fixed down into the structure.

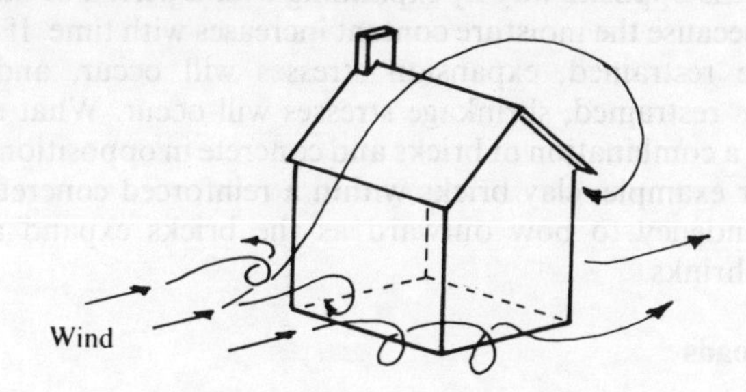

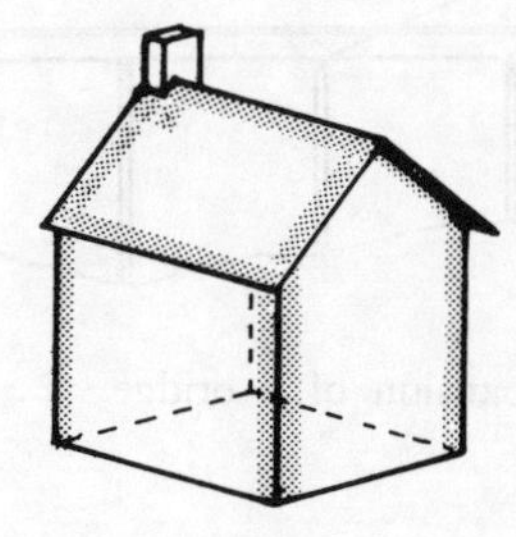

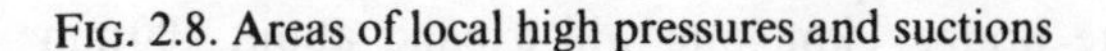

FIG. 2.8. Areas of local high pressures and suctions

2.6 SECONDARY LOADS

The designer should be well aware of secondary loads and how they may affect the structure. What is required is an understanding and an appreciation of when it is prudent to call upon expert advice. The secondary loads are:

1. Shrinkage loads
2. Thermal loads
3. Settlement loads
4. Dynamic loads

Shrinkage loads

Over a period of time, certain building materials such as concrete will shrink and in doing so will set up stresses which, if restrained, may cause cracking. Cracking can be prevented by either providing shrinkage joints or reinforcing the concrete with steel reinforcement so that it has enough strength to cope with the shrinkage. The basic reason why concrete shrinks is that over a period of time, the cement paste in the concrete slowly dries out. On the other hand, clay bricks behave in the opposite way by expanding over a period of time. This happens because the moisture content increases with time. If the clay bricks are restrained, expansion stresses will occur, and if the concrete is restrained, shrinkage stresses will occur. What must be avoided is a combination of bricks and concrete in opposition to each other. For example, clay bricks within a reinforced concrete frame have a tendency to bow outward as the bricks expand and the concrete shrinks.

Thermal loads

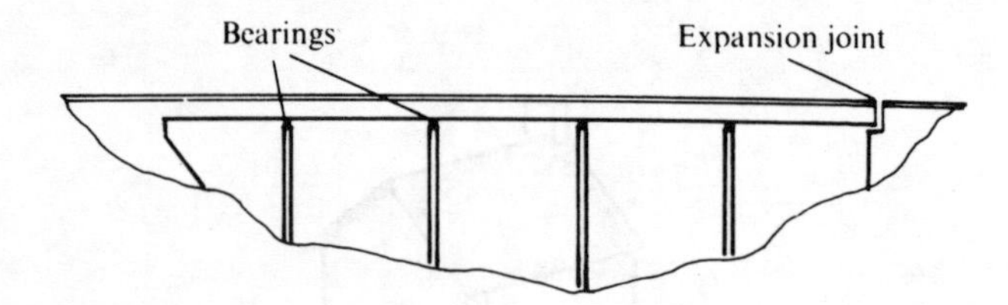

FIG. 2.9. Thermal expansion of a bridge

All building materials expand or contract with temperature change. A concrete bridge 1 km long will expand about 400 mm between winter and summer, and if this expansion was restrained, then enormous thermal loads would be set up causing the bridge to be damaged. This is why bridges have expansion joints and movable bearings so that thermal movement can take place without causing any damage (Fig. 2.9). Long continuous buildings will expand in exactly the same way, and it is necessary to consider the expansion stresses. It is usual to divide a reinforced concrete framed building into lengths not exceeding 30 m and divide a brick wall into lengths not exceeding 10 m. Expansion joints are provided at these points so that the structure is physically separated, and can expand without causing structural damage.

Settlement loads

If one part of a building settles more than another part, then stresses are set up in the structures. If the structure is flexible, then the stresses will be small, but if the structure is stiff, the stresses will be severe unless the two parts of the building are physically separated. A good example is the tower and nave of a church (Fig. 2.10). During

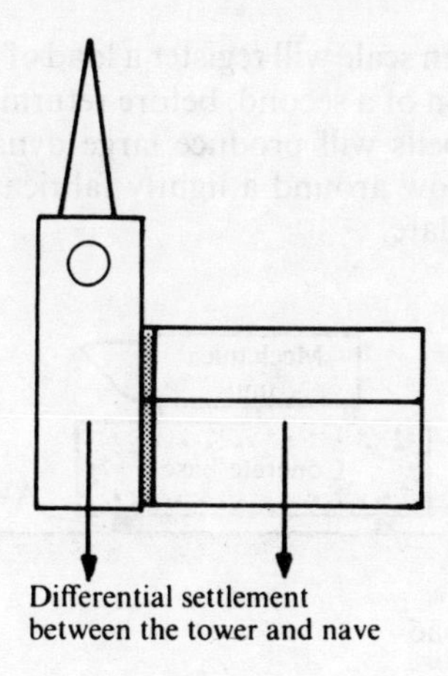

FIG. 2.10. Church

construction, the tower with its extra weight will settle considerably more than the nave. If the tower and the nave are not physically separated, then cracking will occur between the two as differential settlement occurs. A tower block attached to a smaller building will have the same problems, and the two buildings will have to be either physically separated or joined together with a flexible structure (Fig. 2.11).

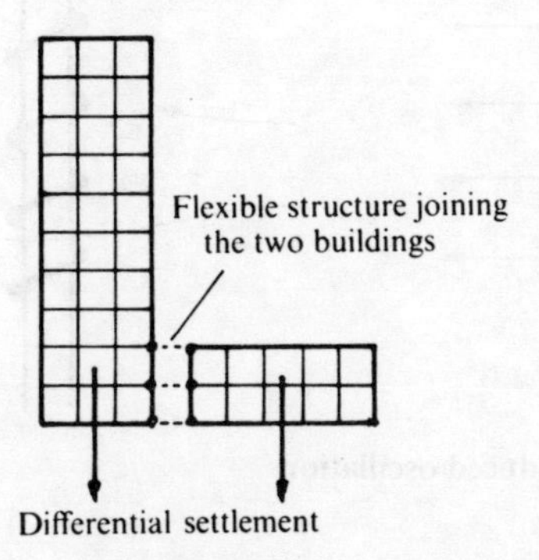

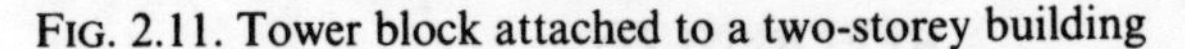

FIG. 2.11. Tower block attached to a two-storey building

Dynamic loads

Dynamic loads, which include impact and aerodynamic loads, are complex. In essence, the magnitude of a load can be greatly increased by its dynamic effect. For example, 1 kg of sugar dropped from a height on to a kitchen scale will register a load of several kilograms on the dial for a fraction of a second, before returning to the 1 kg mark. The resonance of bells will produce large dynamic loads in a bell tower, and the airflow around a lightly fabricated factory chimney will cause it to oscillate.

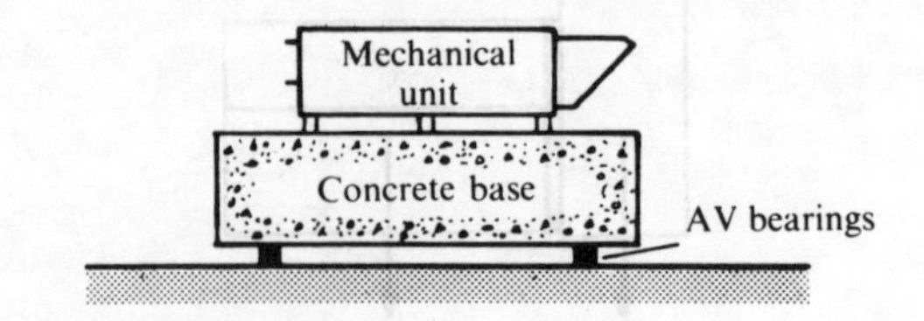

FIG. 2.12. Dynamic load

The solutions to dynamic loads are varied, but in essence the general aim is to find ways of reducing the dynamic effect. An air-conditioning plant placed on a roof can be mounted on AV pads (Fig. 2.12). The AV stands for anti-vibration. An oscillating chimney can be fitted with a spiral strake which disturbs the harmonic flow of the wind eddies around the chimney (Fig. 2.13). The regular pattern of these eddies are the cause of the oscillations and not the magnitude of the wind. If the pattern is made irregular, the oscillations will stop.

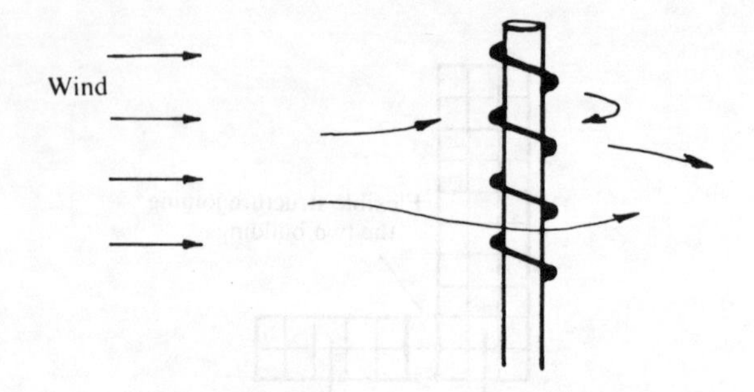

FIG. 2.13. Wind-induced oscillations

REFERENCES

Schedule of Weights of Building Materials, British Standard 648.
Dead and Imposed Loads, British Standard Code of Practice CP3: Chapter V, Part 1.

CHAPTER 3

STRUCTURAL THEORY RELATED TO SIMPLE BEAMS

3.1 FAILURE OF BEAMS DUE TO BENDING, SHEAR AND DEFLECTION

In the previous chapter, strength, stability and serviceability were discussed in relation to the whole building. The structural elements within a building also have to meet these requirements and this section looks at a typical beam (Fig. 3.1) and how it meets these requirements.

FIG. 3.1. Simply supported beam

The beam will have to be strong enough to withstand the loads, stable enough not to fall over, and be stiff enough to meet the function requirement of serviceability. The stability of the beam can be achieved by giving it sufficient lateral bracing, which will hold it firmly in position, and the serviceability or function can be achieved by limiting the amount of deflection. However, the strength will depend upon two forces, the force which will cause the beam to bend and the force which will cause the fibres to shear past each other. The first force could produce failure due to bending and the second, failure due to shear. Assuming then, there is sufficient stability, it is worth looking at a simple beam, and seeing how it may fail under bending, shear and deflection.

Failure due to bending

Consider the beam in Fig. 3.1, which is supported at points A and B. If a number of loads are placed on the beam, the beam will bend and deflect as shown in Fig. 3.2.

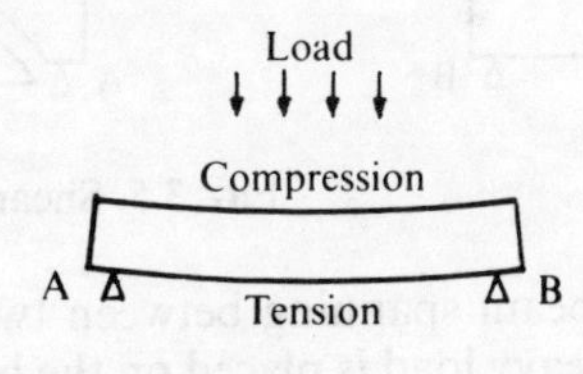

FIG. 3.2

The fibres in the top of the beam will be in compression while the bottom fibres will be in tension. If the tension and compression exceed the natural strength of the material from which the beam has been made, then the beam will fail due to excessive bending stresses.

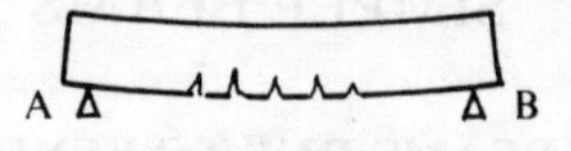

FIG. 3.3. Bending failure

Let us take the example of a concrete beam (Fig. 3.3). The load on the beam will cause tension cracks in the bottom fibres. If the load is increased, the cracks will become larger and the beam will fail. This is because concrete, although very strong in compression, has very little strength in tension, and cannot on its own withstand tension stresses. Where tension occurs, the concrete has to be reinforced with steel reinforcing bars. If the steel bars were placed in the bottom of the beam, then the tension stress will be resisted by the steel and the beam will carry a much greater load before it fails under bending. For failure to occur, either the concrete in the top of the beam has to crush, or the steel in the bottom has to be pulled apart. Which will fail first will depend upon the amount of steel in the bottom of the beam. If there is only a small amount of steel, then the steel will fail first, but if there is a large amount, then the concrete in the top will crush long before the steel will begin to yield.

Some materials such as timber and steel are homogeneous in their structural behaviour in that their maximum compressive strength is about the same as their maximum tensile strength. Bending failure of a beam made from these materials may be due to compression failure in the top of the beam or tension failure in the bottom of the beam, and should occur at the same time.

Failure due to shear

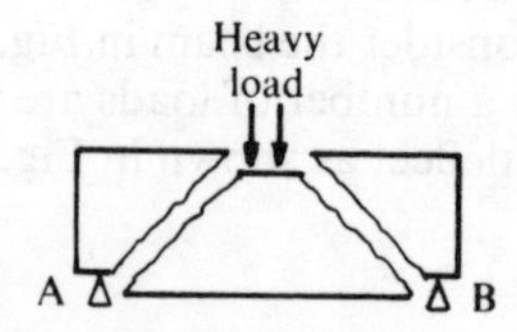

FIG. 3.4. Deep beam

FIG. 3.5. Shear failure

Consider a deep beam spanning between two supports A and B (Fig. 3.4). If a very heavy load is placed on the beam, it will 'punch' a section out of the beam (Fig. 3.5). This kind of failure is called shear failure, because the molecules on one side of the shear line have sheared past the molecules on the other side of the line. The angle of the shear line is usually at about 45 degrees as shown in Fig. 3.5.

If we take the example of a reinforced concrete beam again, we can consider how this beam could be made strong enough to resist the shear forces. Remember that concrete has little strength in tension, so to resist the shear, steel reinforcement has to be placed across the 45-degree 'shear cracks'. This can be done in two ways. The normal way is to use stirrups or links fixed vertically along the beam. These links are vertical reinforcement bars which wrap around and hold the main reinforcement bars in position. Because these links are vertical, they cross over the shear line and prevent the concrete from cracking.

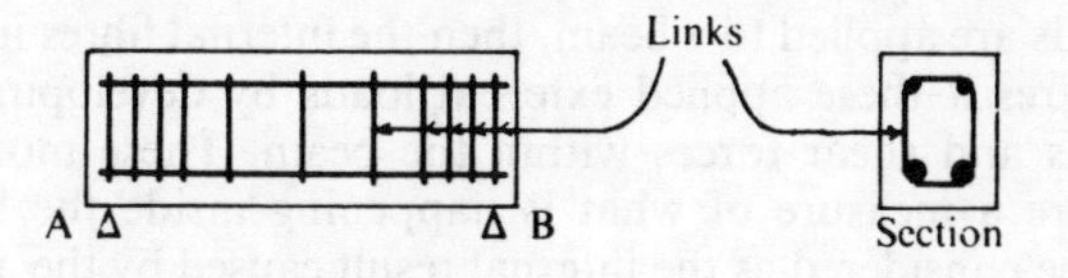

FIG. 3.6. Shear reinforcement links

If the shear loads are very large, it may be necessary to increase the shear strength by bending the main reinforcing bars up at 45 degrees so that they cross the 'shear cracks' at right angles (Fig. 3.7).

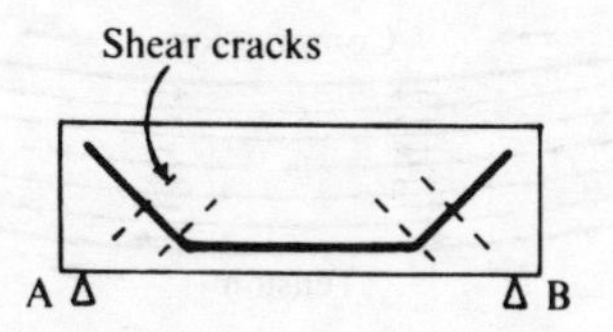

FIG. 3.7. Shear reinforcement bent-up bars

Failure due to deflection

Consider a shallow beam spanning between two supports A and B (Fig. 3.8). A small load on this beam will cause it to deflect. If the load is increased then the deflection may be excessive, causing the beam to look unsafe, although structurally it is not. In addition it may be causing damage to the finishings such as the plaster around the beam, and causing the floors to tilt. Therefore, from a serviceability or function point of view, the beam has failed due to excessive deflection (Fig. 3.9).

FIG. 3.8. Simply supported beam

FIG. 3.9. Deflection failure

3.2 BENDING MOMENT AND SHEAR FORCE

If loads are applied to a beam, then the internal fibres in the beam have to resist these applied external loads by developing bending moments and shear forces within the beam. These moments and forces are a measure of what is happening inside the beam, and should be considered as the internal result caused by the application of external loads.

Bending moment

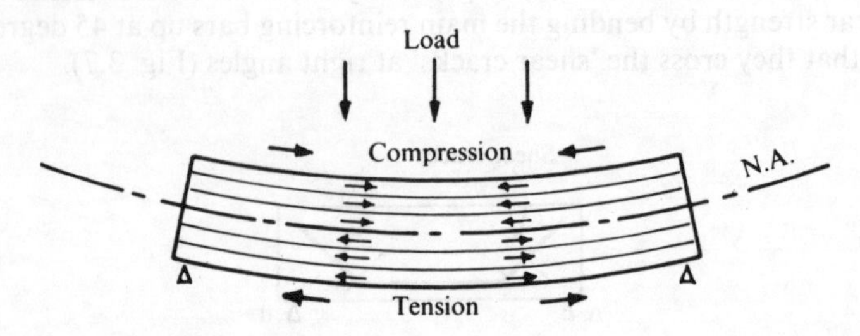

FIG. 3.10. Moments in a beam

When a beam bends under load (Fig. 3.10), the horizontal fibres will change in length. The top fibres will become shorter and the bottom fibres will become longer. That is, the very top fibre will become the shortest and be under maximum compression, while the very bottom fibre will be the longest and be under maximum tension. From the top fibre to the central fibre, the compression gradually decreases until it is zero at the centre, which is called the neutral axis (N.A.). From the neutral axis to the bottom fibre, the fibres are in tension, gradually increasing from zero to a maximum at the bottom fibre.

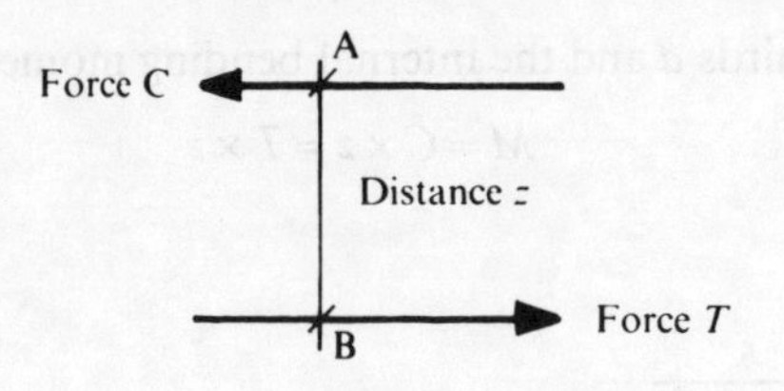

FIG. 3.11

The load on the beam is resisted by the compression force in the top and the tension force in the bottom of the beam. These forces form a 'couple', and this couple produces a moment to counterbalance the external bending moment, which in turn is generated by the applied loads. Now remember that a moment is a force times a distance, and a couple is two equal but opposed forces parallel to one another. A couple cannot be resolved into one force, but it does produce a constant moment about any axis, perpendicular to the plane containing the forces. Thus, in Fig. 3.11, the tension force and the compression force act as a couple, so that:

$$\text{Force } C = \text{Force } T$$
$$\text{Moment about A} = \text{Force } C \times Z$$
$$\text{Moment about B} = \text{Force } T \times Z$$

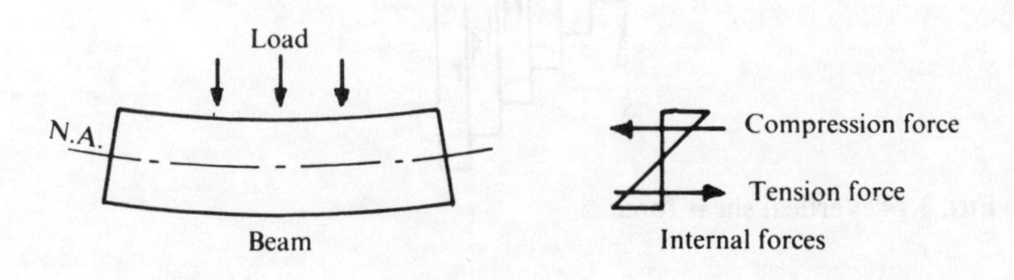

FIG. 3.12

Thus the internal bending moment within a beam (Fig. 3.12) is equal to the distance between the compression and tension forces times either the tension or compression force T or C. The forces are the summation of all the small forces in each fibre varying from zero at the centre to a maximum value at the top and bottom. These small forces can be diagrammatically represented by a triangle as shown in Fig. 3.12 and the total forces will act at the centroid of these triangles. From basic geometry we know that the centroid of a triangle is one-third the height above the base, and it can be seen from Fig. 3.13, that if the depth of the beam is d, then the distance between the forces equals two-thirds d and the internal bending moment equals M:

$$M = C \times z = T \times z$$

where $z = \frac{2}{3}d$.

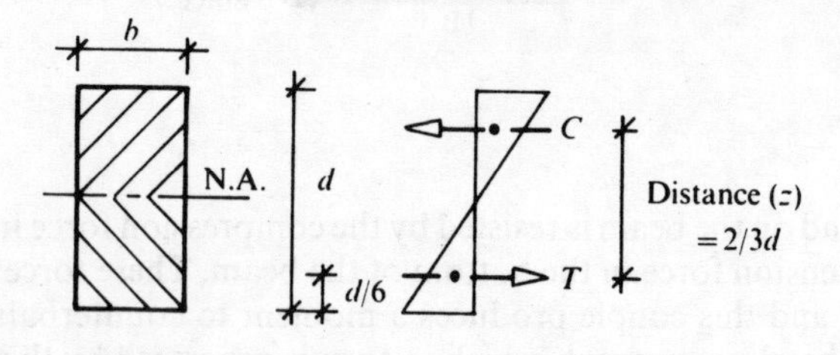

FIG. 3.13

Shear force

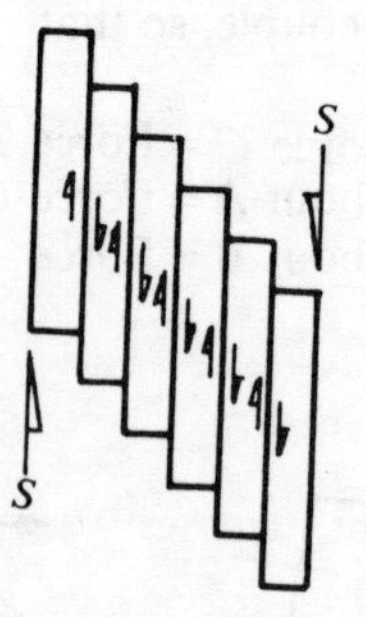

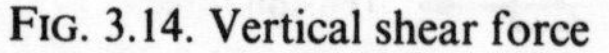
FIG. 3.14. Vertical shear force

A small section of a beam is shown in Fig. 3.14. On the left there is a vertical force pushing upwards, and on the right there is a vertical force pushing downwards. The fibres between the two forces will try to slide or shear over one another. These forces are described as the vertical shear forces and are generated by the loads placed on the beam and the corresponding support reactions, the reactions pushing upwards and the loads pushing downwards.

The values of the shear force and bending moment will vary along the beam. As the beam has to be made large enough and strong enough to resist the maximum shear force and bending moment, it is useful to plot their values, so that the positions of the maximum values can be located. The diagram produced by the plottings are called shear-force and bending-moment diagrams, and they are a direct reflection of the ideal visual proportions for the beam. This is because the depth of the beam at any section is a function of the bending moment and the shear force. If the shear force is relatively small, then the depth should follow the pattern of the bending moment. This is very evident in the design of bridges and long span roofs (Fig. 3.15), but it is not often very practical to vary the depth of a beam within a building.

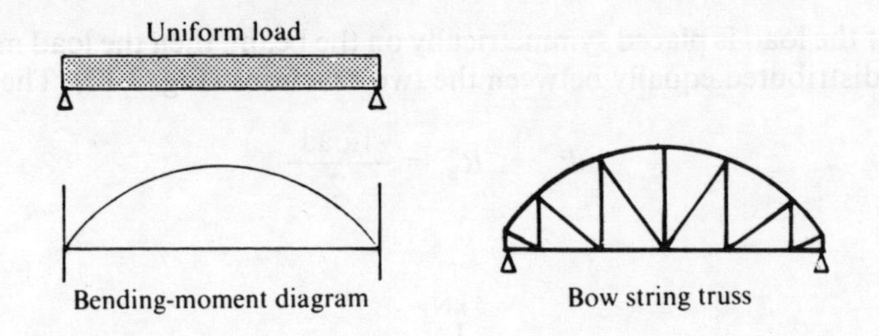

FIG. 3.15

3.3 BENDING-MOMENT AND SHEAR-FORCE DIAGRAMS

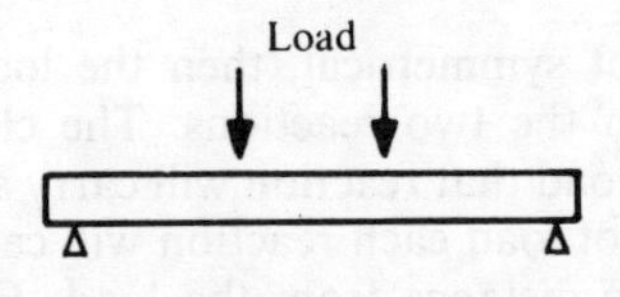

FIG. 3.16

Consider a simply supported beam AB with a load in the middle (Fig. 3.16). Before it is possible to calculate and plot the bending-moment and shear-force diagrams, it is necessary to calculate the beam reactions R_A and R_B. The reactions are the static forces exerted by the supports on the beam to balance the downward force of the applied load, so the greater the load, the greater the reactions will be.

Calculation of the reactions

The beam reactions must support the applied load and as the beam is in equilibrium, the vertical downward load must equal the upward vertical reactions (Fig. 3.16). Therefore:

$$R_A + R_B = \text{Vertical load}$$

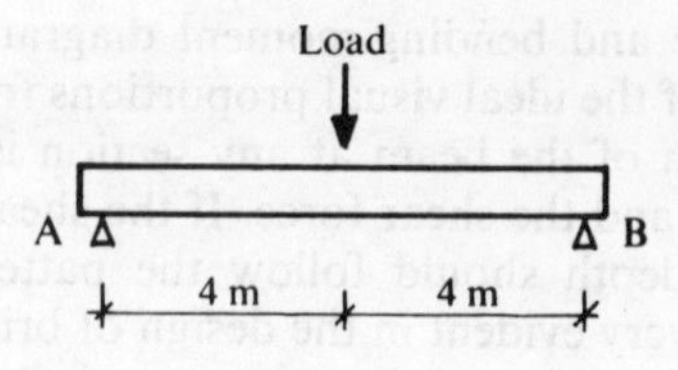

FIG. 3.17

If the load is placed symmetrically on the beam, then the load must be distributed equally between the two reactions (Fig. 3.17). Then

$$R_A = R_B = \frac{\text{Load}}{2}$$

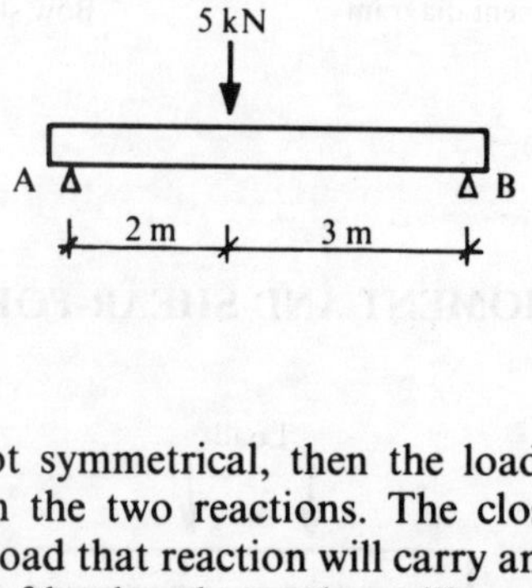

5 kN
A
B
2 m
3 m

FIG. 3.18

If the load is not symmetrical, then the load will be unevenly distributed between the two reactions. The closer a load is to a reaction, the more load that reaction will carry and in mathematical terms, the amount of load each reaction will carry will be inversely proportional to the distance from the load. Figure 3.18 shows a simply supported beam AB with an unsymmetrical load of 5kN. The load is closer to A than B, so R_A will carry more load than R_B. Using the inversely proportional rule:

$$R_A = \frac{3\,\text{m}}{5\,\text{m}} \times 5\,\text{kN} = 3\,\text{kN}$$

$$R_B = \frac{2\,\text{m}}{5\,\text{m}} \times 5\,\text{kN} = 2\,\text{kN}$$

Another way of calculating the reactions is to take moments about points A and B as well as resolving vertically.
Resolving the forces vertically:

$$R_A + R_B = 5\,\text{kN}$$

Take moments about B:

$$\text{Clockwise moments} = \text{Anticlockwise moments}$$
$$5\,\text{m} \times R_A = 3\,\text{m} \times 5\,\text{kN}$$
$$R_A = 3\,\text{m}/5\,\text{m} \times 5\,\text{kN} = 3\,\text{kN}$$

Take moments about A:

$$\text{Clockwise moments} = \text{Anticlockwise moments}$$
$$2\,\text{m} \times 5\,\text{kN} = 5\,\text{m} \times R_A$$
$$R_B = 2\,\text{m}/5\,\text{m} \times 5\,\text{kN} = 2\,\text{kN}$$

Calculations for the shear-force and bending-moment diagrams

There are a number of ways to plot the shear-force and bending-moment diagrams. A simple way is to use a paper overlay as shown in Fig. 3.19. The left-hand edge of the overlay will be the point where the moment and shear force is being calculated. It is important to appreciate that this point moves when the overlay is moved across the beam, so that each time the overlay is moved, a new point is being considered which will have a different bending moment and shear force.

If we start by placing the overlay to the extreme left so that only the reaction R_A is visible, then the shear force will be equal to the sum of all the forces visible to the left of the overlay. In this case, the only force visible is R_A. Therefore:

Shear force $= R_A = 3$ kN
Bending moment $= R_A \times$ the distance from the edge of the overlay to the force R_A
$= R_A \times 0 = 0$

The overlay can now be moved to 1 m from the end of the beam and the process repeated. Again the shear force will be equal to the sum of all the forces visible to the left of the overlay, and again the only force visible is R_A. But this time the moment will not be zero, as the distance from the left-hand edge of the overlay to the reaction force is now 1 m. At 1 m:

$$\text{Shear force} = R_A = 3\,\text{kN}$$
$$\text{Bending moment} = R_A \times 1\,\text{m}$$
$$= 3\,\text{kN} \times 1\,\text{m} = 3\,\text{kN·m}$$

If the overlay is now moved to 2 m, the only force visible is still R_A. At 2 m:

$$\text{Shear force} = R_A = 3\,\text{kN}$$
$$\text{Bending moment} = R_A \times 2\,\text{m}$$
$$= 3\,\text{kN} \times 2\,\text{m} = 6\,\text{kN·m}$$

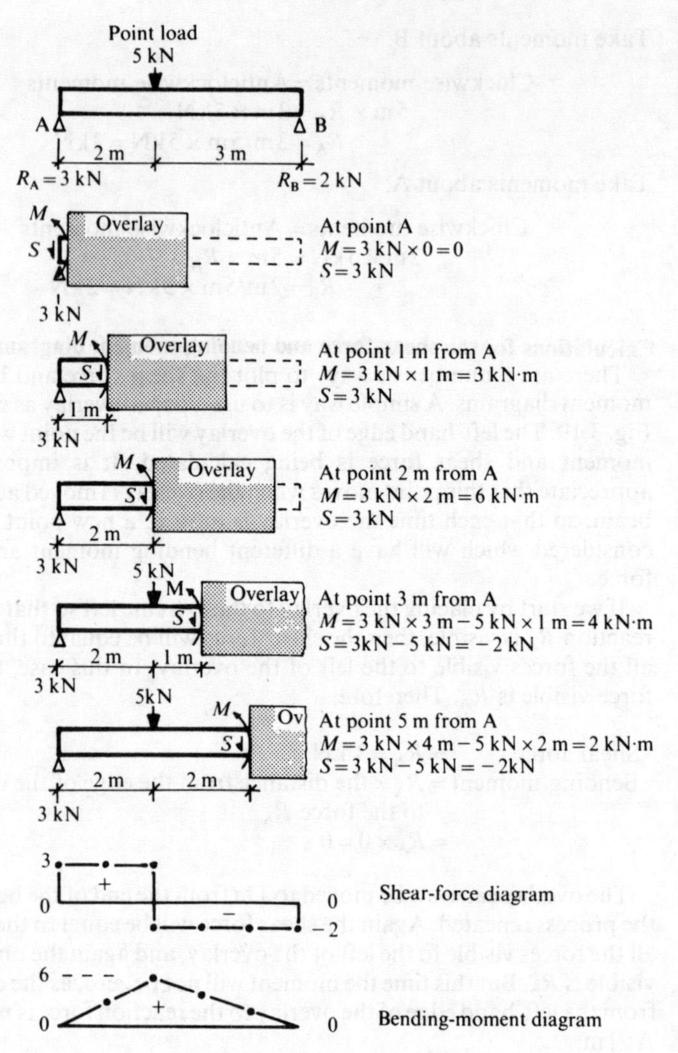

FIG. 3.19. Calculation of shear forces and bending moments

At 3 m there are now two vertical forces. The reaction in an upward direction and the load in a downward direction. The reaction will produce a clockwise moment while the load will produce an anticlockwise moment.

At 3 m:

$$\begin{aligned} \text{Shear force} &= R_A - \text{Load} \\ &= 3\,\text{kN} - 5\,\text{kN} = -2\,\text{kN} \\ \text{Bending moment} &= R_A \times 3\,\text{m} - \text{Load} \times 1\,\text{m} \\ &= 3\,\text{kN} \times 3\,\text{m} - 5\,\text{kN} \times 1\,\text{m} \\ &= 9\,\text{kN·m} - 5\,\text{kN·m} = 4\,\text{kN·m} \end{aligned}$$

The process is repeated along the beam at sufficient intervals so that the shear-force and bending-moment diagrams can be drawn (Fig. 3.19).

Sign conventions

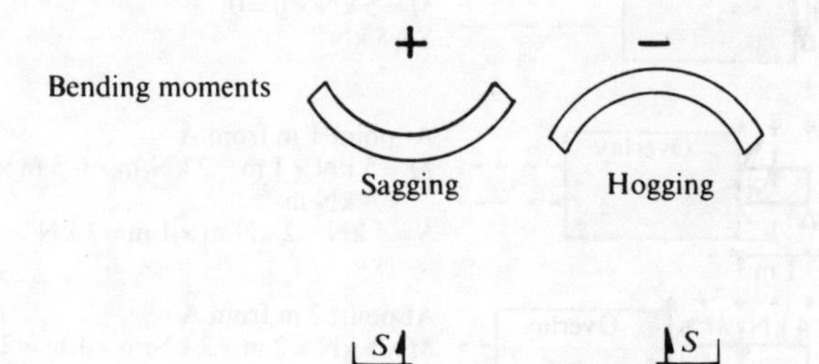

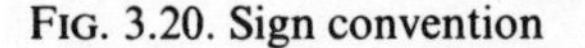
FIG. 3.20. Sign convention

Whether a bending moment or shear force is described as positive or negative will depend upon an arbitrary sign convention. The normal sign convention and the one used in this book is given in Fig. 3.20. A bending moment which causes the beam to sag is called positive and one which causes it to hog, negative. The sign convention for shear force is positive when the left-hand side of the beam is being pushed upwards and the right-hand side is pushing downwards. The shear force is negative when the forces are reversed.

Uniformly distributed loads

Most building structures are subjected to loads which are evenly distributed along the beams or over the floor slabs. The bending-moment and shear-force diagrams for this kind of load can be plotted in the same way as for the point load (Fig. 3.19). The overlay is placed over the beam so that the shear force will equal the sum of all the visible forces, while the bending moment will equal the sum of all the moments about the edge of the overlay (Fig. 3.21).

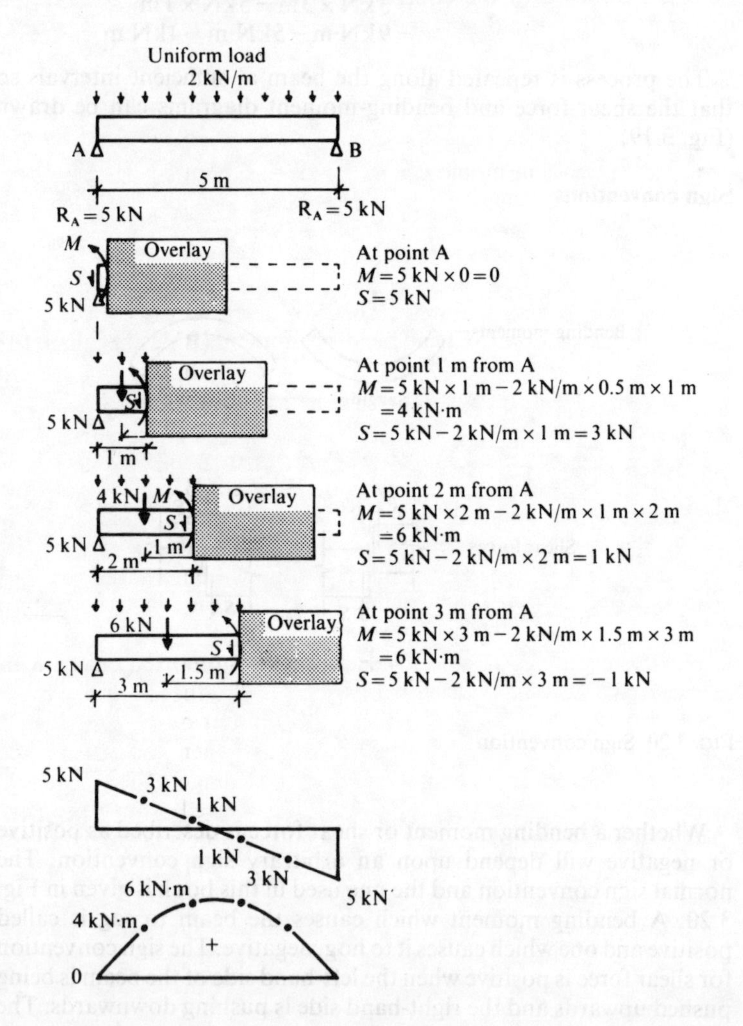

FIG. 3.21. Calculation of shear force and bending moments

At each position of the overlay, the uniformly distributed load visible will be equivalent to a point load acting at its centre. The bending moment at the point where the overlay cuts the beam will be the difference between the moment due to the reaction and the moment due to the equivalent point load. The reaction will be pushing the beam upwards, and the load will be pushing it downwards, so that is why the bending moment will be the difference between the two. If the distance to the reaction from the overlay is x, then the distance to the equivalent point load will be $x/2$. If the uniform load on the beam is w kN for every metre length of beam, then:

$$\text{Shear force} = R_A - w \cdot x \quad \text{kN}$$

$$\text{Bending moment} = R_A - w \cdot x \cdot \frac{x}{2} \quad \text{kN·m}$$

These are the general equations for the shear force and bending moments. Figure 3.21 shows a simply supported beam AB with a uniform load of 2 kN per metre length of beam. Using the equation above, the shear force (SF) and bending moment (BM) are calculated for each metre length across the beam by using the overlay at each point. The SF and BM diagrams are then plotted so that the SF is a straight line varying from + 5 kN to 45 kN, and the BM is a parabolic curve with the maximum moment in the centre.

It should be noted that in both Figs 3.20 and 3.21, and with the worked examples that follow, the BM is a maximum when the SF is zero. If the position of the maximum BM is difficult to locate, then its position can be calculated by finding the position where the SF is zero. The other observation worth noting is that the SF is a maximum at the supports.

Worked examples 3.3 and 3.4 show simply supported beams with cantilevers. The effect of the cantilevers is to produce 'hogging' over the supports, which in turn reduces the amount of sagging at mid span. If the amount of sagging is reduced, then the size of the structural member can be reduced, thus providing a more economic structure. Referring to the second of these worked examples again, the maximum moment is 13.5 kN·m. If both cantilevers are removed, then this value will be increased to 18 kN·m (13.5 + 4.5), which illustrates the increased efficiency of the beam with the cantilevers. This principle can be extended to a continuously supported beam or slab (Fig. 3.22) where the hogging moments at the supports reduce the sagging moments at mid span. Therefore, a beam continuous over several supports can have a greater span to depth ratio than a simply supported beam.

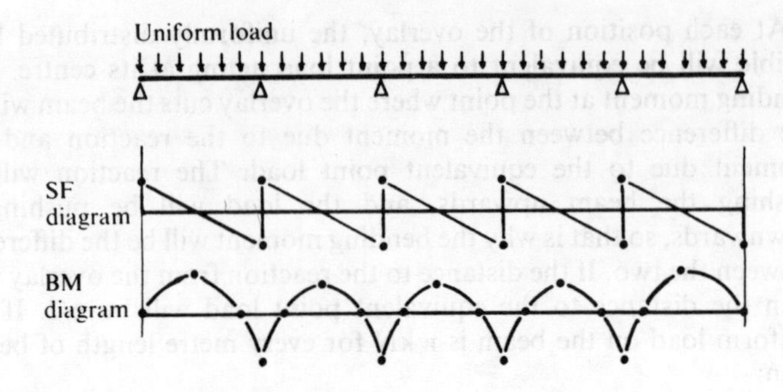

FIG. 3.22. Continuously supported beam

3.4 LAWS OF BENDING

In the previous sections it was shown that when a simply supported beam deflects under load, the top fibres are in compression and the bottom fibres are in tension. Somewhere in between, there is the neutral axis where the fibres are neither in tension nor compression. At this point the stress is zero. Moving away from the neutral axis, the stress gradually increases until it is at a maximum compressive stress in the very top fibre and a maximum tensile stress in the very bottom fibre (Figs 3.23 and 3.24). To calculate the required beam size, it is necessary to equate these maximum stresses with the applied loads and BMs. The relationships between the stresses, the BMs, the stiffness of the beam section and the materials used, are called the 'Laws of Bending'.

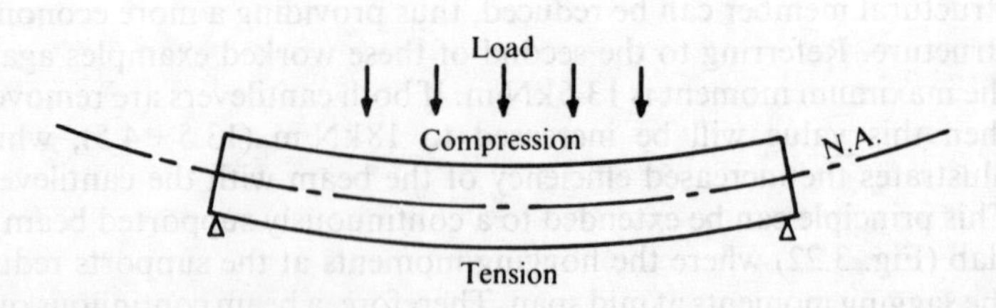

FIG. 3.23. Bending stresses

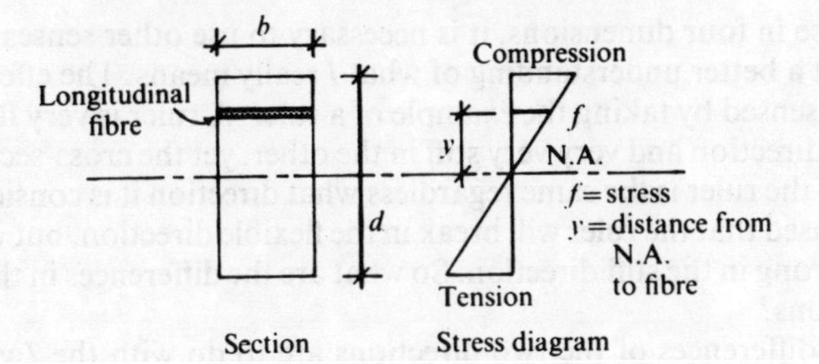

FIG. 3.24. Section through beam

The Laws

$$\text{Stress} = f = \frac{M}{I}\, y$$

OR

$$\text{Stress} = f = \frac{M}{Z}$$

where f = stress
M = bending moment
I = second moment of area or moment of inertia (stiffness of the section)
y = distance from the neutral axis
Z = section modulus = I/y or elastic modulus

If the reader wishes to know how the Laws of Bending are derived, then they are advised to refer to a book on the theory of engineering structures as the derivation of this formula is beyond the scope of this book.

Moment of inertia or second moment of area

The stress (f) and bending moment (M) have been dealt with in previous sections. The I and Z are new concepts and difficult ones to understand at first. The I is sometimes called the moment of inertia and at other times called the second moment of area, but their meaning is the same. Steel design manuals use the term 'moment of inertia', while timber handbooks use 'second moment of area'.

Not only is it the confusion of the name which makes I difficult to comprehend, but it is the fact that it has fourth-dimensional properties, its units being mm^4. As it is not usual to think and visualise in four dimensions, it is necessary to use other senses to try and get a better understanding of what I really means. The effect of I can be sensed by taking the example of a ruler. A ruler is very flexible in one direction and very very stiff in the other, yet the cross-sectional area of the ruler is the same regardless what direction it is considered. It is sensed that the ruler will break in the flexible direction, but will be very strong in the stiff direction. So what are the differences in the two directions?

The differences of the two directions are to do with the I values, which are very different in the two directions. This is illustrated in mathematical terms by comparing the values of I in the two directions for a rectangular section in Fig. 3.25. I is equal to $bd^3/12$, where b is the breadth of the section and d is the depth. I is then proportional to d^3 and for the example of the ruler, d is small in one direction compared with d in the other direction, so the difference is considerably magnified when the cube of d is taken. This is best demonstrated by giving the ruler some dimensions.

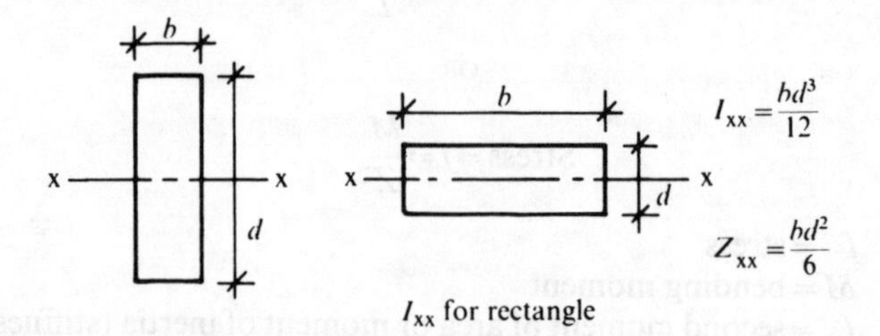

FIG. 3.25. Moment of inertia (stiffness of beam)

Let the cross-sectional dimensions of the ruler be 4 mm × 30 mm. Then:

$$\text{when } d = 30 \text{ mm } I\frac{b \times d^3}{12} = \frac{4 \times 30 \times 30 \times 30}{12} = 9000 \text{ mm}^4$$

$$\text{and } d = 4 \text{ mm } I = \frac{b \times d^3}{12} = \frac{30 \times 4 \times 4 \times 4}{12} = 160 \text{ mm}^4$$

This means that the ruler is approximately 56 times stiffer in the stiff direction than it is in the flexible direction.

Section modulus or elastic modulus

The section modulus (Z), being a derivative of I, is also a measure of stiffness. When calculating the size of a structural beam, it is usual to consider the worst possible conditions of stress. As the maximum bending stresses occur on the outer fibres of a beam, then the distance from the neutral axis to the maximum stress will be equal to half the depth for a symmetrical beam. The Laws of Bending can, therefore, be simplified to $f = M/Z$ by putting $y = d/2$ and $I/y = Z$. So that:

$$\text{Section modulus } Z = \frac{I}{y} = \frac{bd^3/12}{d/2} = \frac{bd^2}{6}$$

The Laws of Bending now become:

$$\text{Max. stress in a beam } (f) = \frac{\text{Max. bending moment } (M)}{\text{Section modulus } (Z)}$$

3.5 CALCULATING BEAM DEFLECTIONS

In section 1.2 it was shown how important it is to limit deflections in building structures so that the function or serviceability of the building is not impeded. In timber design, it is recommended that the deflection of beams is limited to 0.003 of the span, while in steel design a value of 1/360 of the span is recommended. In practice, these values are very similar and can be considered to be the same.

The deflection is dependent on all the following factors:

1. Load
2. Span
3. Stiffness of the material (E)
4. Stiffness of the beam section (I)

That is, the greater the load the larger the deflection, the greater the span the larger the deflection, but the greater the stiffness of the material (E) and the beam section (I) the smaller the deflection will be. Therefore:

$$\text{Deflection will vary as } \frac{\text{load} \times (\text{span})^3}{E \times I}$$

OR

$$\text{Deflection} = \text{constant} \times \frac{WL^3}{EI}$$

where W = load
L = span
E = Young's modulus
I = second moment of area

The value of the constant will depend upon the support and loading conditions of the beam. The values of the constants are given in Appendix 4 for a number of loading conditions. For example, for a uniform load on a simply supported beam, the constant equals 5/384.

Worked Examples 3.1 and 3.2
Project: Shear force and bending-moment diagrams
Point loads

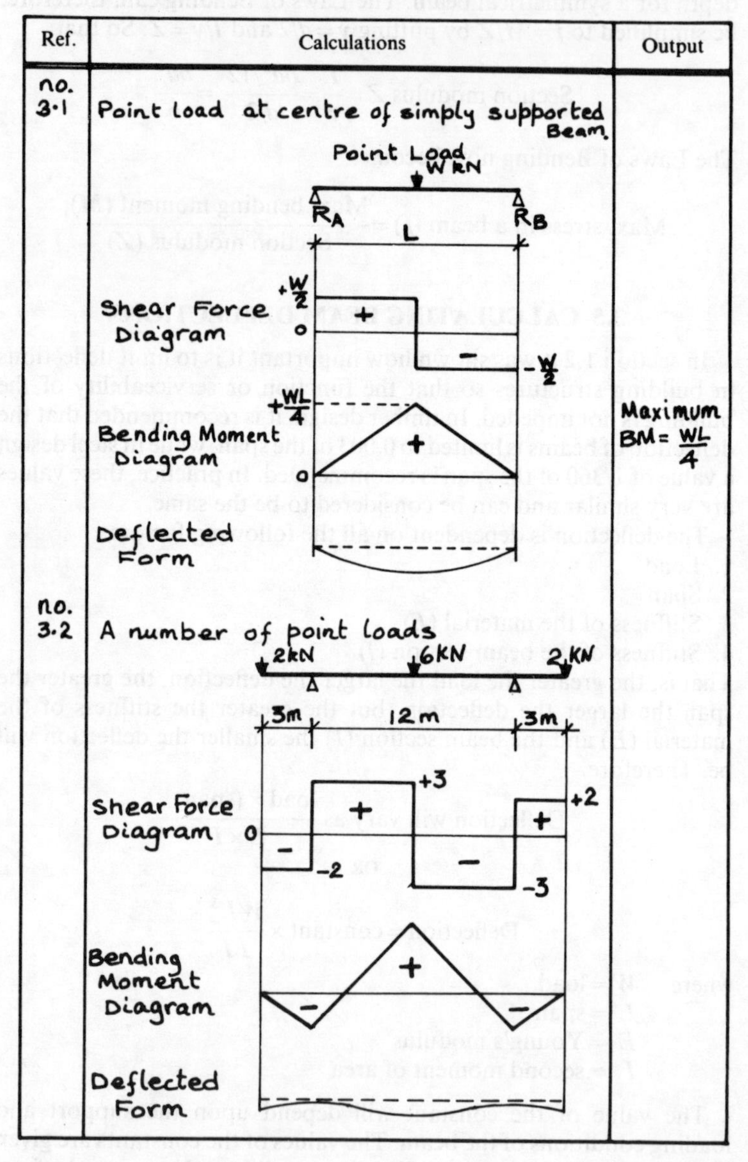

Worked Examples 3.3 and 3.4
Project: Shear force and bending-moment diagrams
Uniformly distributed loads

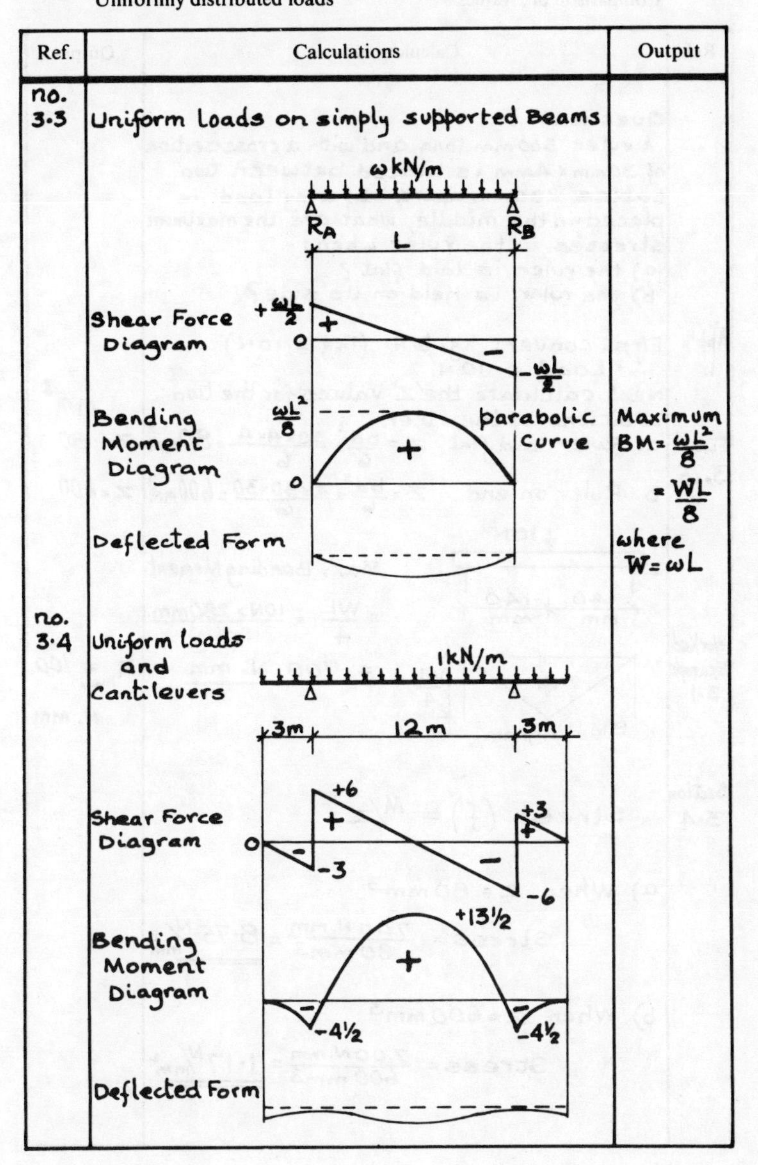

Worked Example 3.5
Project: The stiffness of a 'ruler'
Comparison of z values

Ref.	Calculations	Output
	Question A ruler 300 mm long and with a cross-section of 30 mm × 4 mm, is placed between two tables 280 mm apart. If a 1 kg load is placed in the middle, what are the maximum stresses in the ruler when:- a) the ruler is laid flat? b) the ruler is held on its side?	
App'x 1.	First convert kg's to N (1 kg = 10 N) ∴ Load = 10 N.	
	Next calculate the 'Z' values for the two positions of the ruler.	mm^3
Fig. 3.25	a. Ruler laid flat $Z = \frac{bd^2}{6} = \frac{30 \times 4 \times 4}{6} = 80\,mm^3$	$Z = 80$
	b. Ruler on end $Z = \frac{bd^2}{6} = \frac{4 \times 30 \times 30}{6} = 600\,mm^3$	$Z = 600$
	↓10 N; 140 mm, 140 mm. Max. Bending Moment $= \frac{WL}{4} = \frac{10\,N \times 280\,mm}{4}$	
Worked Example 3.1	BM diagram (+, $\frac{WL}{4}$) $= 700$ N. mm	$M_{max} = 700$ N. mm
Section 3.4	Stress $(f) = M/Z$	
	a) When $Z = 80\,mm^3$ Stress $= \frac{700\,N.mm}{80\,mm^3} = 8{\cdot}75\,N/mm^2$	
	b) When $Z = 600\,mm^3$ Stress $= \frac{700\,N.mm}{600\,mm^3} = 1{\cdot}17\,N/mm^2$	

Worked Example 3.6
Project: Deflection of a 'ruler'

Ref.	Calculations	Output
	Question Find the deflection of the ruler in worked example 3.5 when the ruler is placed on its side.	
	10N. (10N = 1kg) 280 mm	
	Load W = 10N Length L = 280 mm Young's modulus E = 8KN/mm² second moment of area $I = bd^3/12$	
	The value taken for 'E' assumes the ruler is made from softwood.	
App'x 5	Second moment of area 'I' $I = \frac{bd^3}{12} = \frac{30\,mm \times 4\,mm \times 4\,mm \times 4\,mm}{12}$ $= 160\ mm^4$	$I = 160\ mm^4$
App'x 4	Deflection for a point load $= \frac{1}{48} \times \frac{WL^3}{EI}$ $= \frac{1}{48} \times \frac{10N \times 280\,mm \times 280\,mm \times 280\,mm}{8\,000\,N/mm^2 \times 160\,mm^4}$ $= 3.6\ mm$	Deflection = 3.6mm

CHAPTER 4

BASIC STRUCTURAL THEORY RELATED TO TRUSSES

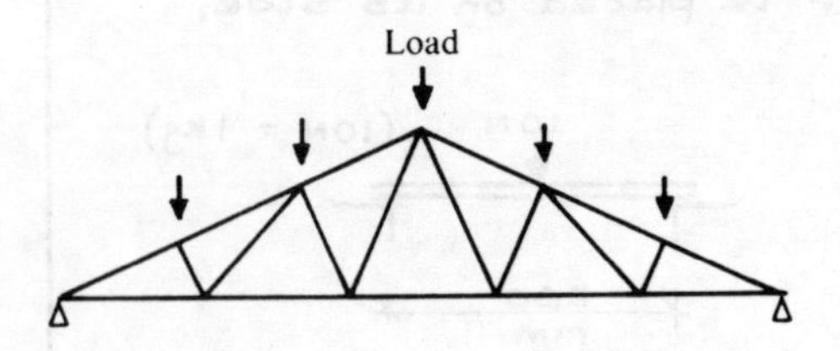

FIG. 4.1. Roof truss

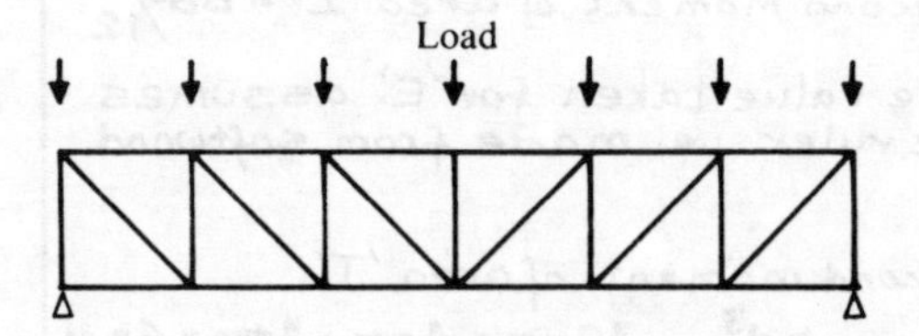

FIG. 4.2. Truss

Trusses can be many different shapes and sizes, but generally they can be divided into either triangular forms which are ideally suited for roofs as shown in Fig. 4.1, or truss types that are really an extension of a beam to carry loads using less material but with the disadvantage of a deeper section (Fig. 4.2).

The reduction in weight is achieved by eliminating all the BMs. The general principle that runs through all structural forms, is that the structure becomes much more efficient, if BMs are eliminated and the loads are carried instead by structural members in direct compression or tension. Hence, the difference between a truss and a beam is that the structural members in a truss are either in tension or compression, and not subjected to bending as in the case of a beam.

For a truss to behave in this way, the joints have to be pin jointed, that is, free to rotate if a small moment is applied to a member. In practice most joints behave in this way even though they may look quite rigid with two or more bolts holding the members to the gusset plate. For a joint to be rigid and be able to transfer a moment, it will have to have a fully welded connection so that no movement can occur at the joint. Trusses connected in this way will not be free of internal BM, and may use more material than a pin-jointed truss.

However, as has been mentioned above, the reduction in the weight is achieved at the expense of the span to depth ratio. If the depth is critical, then a compromise between a beam and a pin-jointed truss can be achieved by having a fully welded truss with members designed to take small bending moments. This kind of truss will give a shallower depth.

4.1 FORCES AND VECTOR COMPONENTS

Parallelogram of forces

A force has magnitude and direction

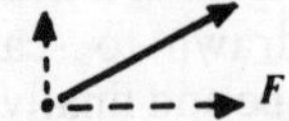

A force can be resolved into other directions

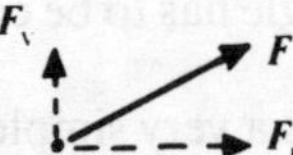

F_V – Vertical component

F_H – Horizontal component

FIG. 4.3. Forces

A force is what the mathematicians call a vector quantity as it has both magnitude and direction (Fig. 4.3). There are certain laws which govern vectors such as the parallelogram law used to compound two vectors together. Figure 4.4 illustrates how the resultant of two forces is calculated by using the parallelogram law.

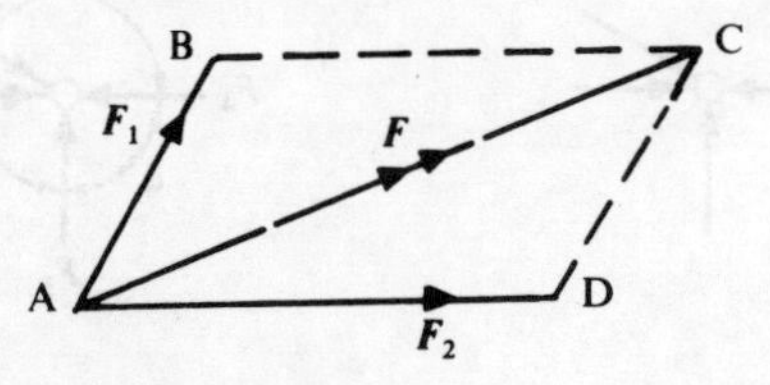

FIG. 4.4. Parallelogram of two forces

If two forces represented by the adjacent sides AB and AD of a parallelogram ABCD act at a point A, then the resultant force is represented by the diagonal AC. If the forces are drawn to scale, then the magnitude of the resultant can be scaled from the diagram. Conversely, a force can be resolved into a vertical and horizontal component (Fig. 4.3). Again the magnitude of F_V and F_H can be scaled from the diagram.

Triangle of forces

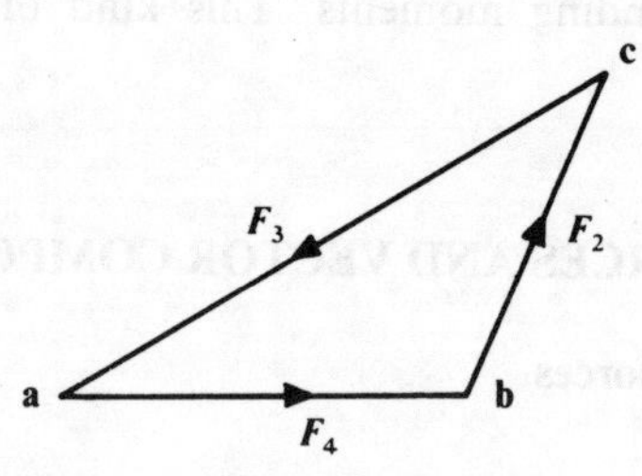

FIG. 4.5. Triangle of forces

If three forces, acting at a point, are in equilibrium they can be represented in magnitude and direction by the sides of a triangle (Fig. 4.5). Starting at a point **a**, the for F_1 is drawn to scale in the direction **ab**. Then F_2 is drawn in the direction **bc** and finally the force F_3 is drawn in the direction **ca**. If the three forces are in equilibrium, force F_3 will finish where F_1 started. The triangle has to be completed for the forces to be in balance.

This concept of a triangle of forces provides a very simple method of determining the magnitude and direction of unknown forces. For example, if the direction of all the forces can be plotted, then provided the magnitude of one of the forces is known, the other two forces can be scaled from the triangle.

Polygon of forces

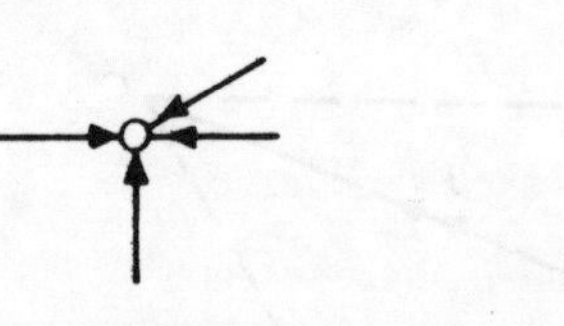

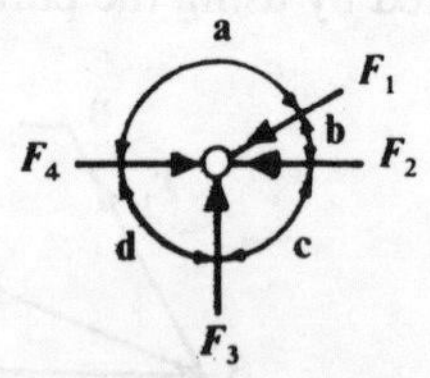

FIG. 4.6. Forces at a point

Where there are more than three forces (Fig. 4.6), the forces can be resolved into a polygon of forces. As with the triangle of forces, the polygon has to be completed for the forces to be balanced and in equilibrium (Fig. 4.7).

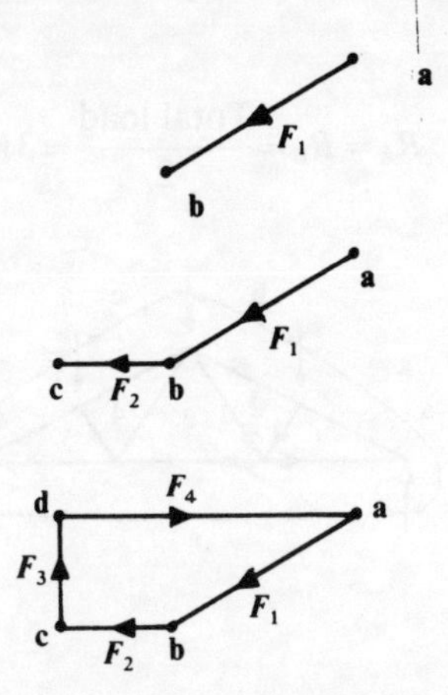

FIG. 4.7

The forces acting at a point are numbered in a clockwise direction, F_1, F_2, F_3, etc., and the spaces in between the forces are lettered, **a**, **b**, **c**, etc. (Fig. 4.6). Now the polygon is drawn starting at point **a** so that **ab** represents the direction and magnitude of the force F_1. The process is continued for **bc**, **cd** and **da**, representing forces F_2, F_3 and F_4. For the system of forces to be in equilibrium the last force must close the polygon at point **a**.

4.2 BOW'S NOTATION

Bow's notation is a graphical method of drawing several polygons of forces on one diagram, and if drawn to scale will enable forces to be measured directly from the diagram. This method is particularly useful when determining the forces in a truss where there are forces acting on a number of joints (Fig. 4.8).

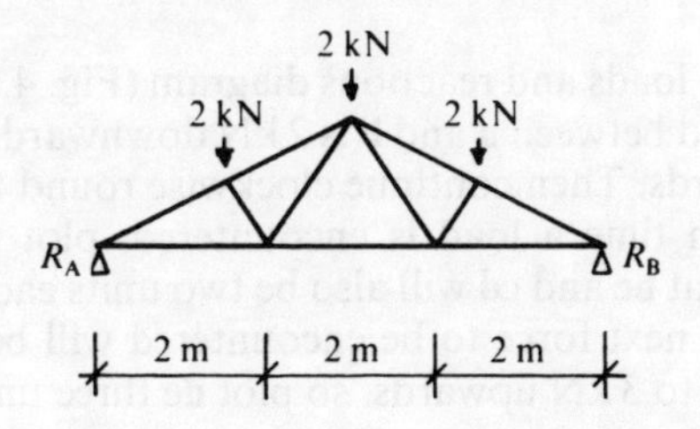

FIG. 4.8

Bow's notation is best explained by using an example. Therefore, consider a roof truss loaded as shown in Fig. 4.8 with three point loads of 2 kN each and a span of 6 m. By inspection, the reactions must be equal as the loads are symmetrically placed on the truss. Therefore:

$$R_A = R_B = \frac{\text{Total load}}{2} = 3\,\text{kN}$$

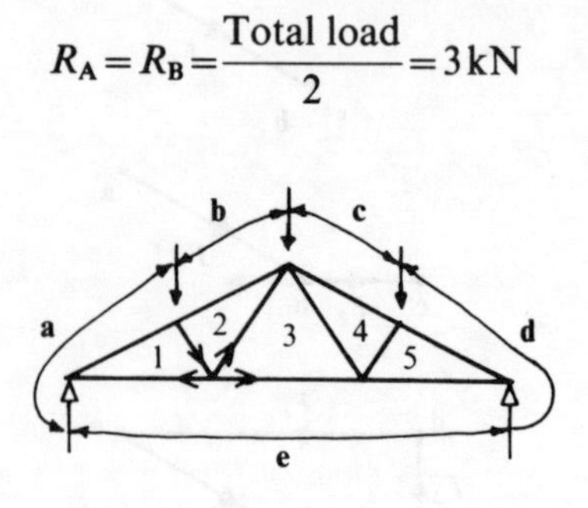

FIG. 4.9

The notation used to identify each member of the truss is shown in Fig. 4.9. This is done by lettering the spaces in between the external loads and reactions, **a**, **b**, **c**, **d**, etc. Now identify the triangular spaces within the truss by using numbers 1, 2, 3, 4, etc., so that every number in the truss can be referred to by the pair of letters or numbers, which appear on either side of that member. For example, **a**1 is the first top member in the truss, and **b**2 the next member.

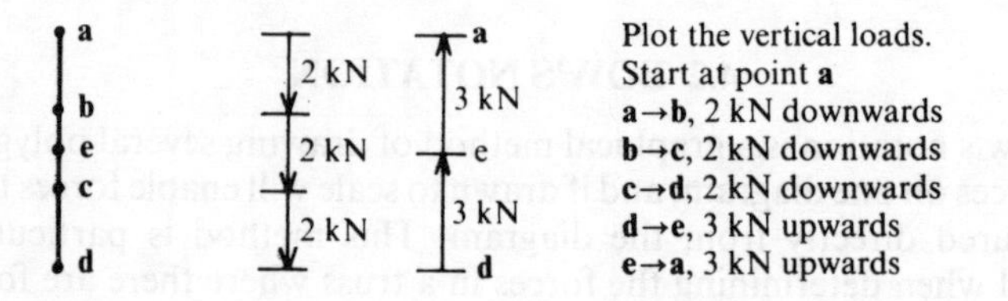

FIG. 4.10. Loads and reaction diagram

To draw the loads and reactions diagram (Fig. 4.10), start at point **a**. The first load between **a** and **b** is 2 kN downwards, so draw **ab** two units downwards. Then continue clockwise round the outside of the truss and each time a load is encountered, plot that load on the diagram, so that **bc** and **cd** will also be two units each in a downward direction. The next force to be encountered will be the reaction R_B which is equal to 3 kN upwards, so plot **de** three units upwards. The last force will be R_A which is also equal to 3kN upwards, and returns the plot on the diagram to **a**. This diagram of loads and reactions is now the base line for Bow's notation. That is, the polygon of forces for each joint in the truss can now be drawn from this base line with each member exerting either a tension or compression force.

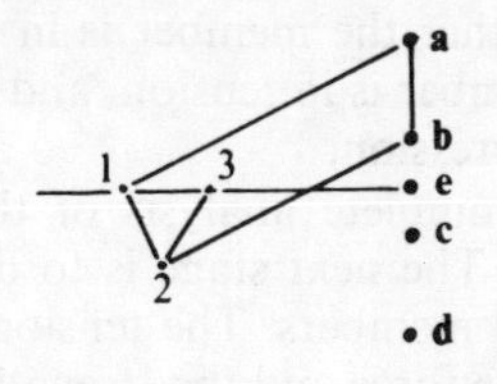

FIG. 4.11. Forces in members

These forces are now represented on the diagram by drawing a line parallel to the member through the appropriate space on the base line. For a member **a**1 a line is drawn through point **a** on the base line and for member **e**1 a line is drawn through **e**. Where these lines meet on the diagram will be point 1, so that the forces in the members **a**1 and **e**1 will be the scaled distances **a**1 and **e**1 on the diagram. This process is continued for the other members in the truss until the Bow's notation diagram is completed (Fig. 4.11), and the results summarised in Table 4.1.

The question arises about whether a member is in tension or compression. For most members this can be decided by inspection. For example, common sense suggests that the top members are all in compression and the bottom members are all in tension. The members in between are more difficult to decide, and it is necessary to go back to the Bow's notation diagram and find the polygon of forces for each particular joint. To illustrate this, take the polygon **ab**21. The forces will flow round the polygon, so starting with the external

TABLE 4.1 Forces in members

Members		*Force (kN)*	
a1	**d**5	6.7	Compression
b2	**c**4	5.6	Compression
e1	**e**5	6.0	Tension
e3		4.0	Tension
12	45	1.8	Compression
23	43	1.8	Tension

load **ab**, which is a downward force, plot directional arrows around the polygon. If the arrows are now plotted on the truss (Fig. 4.9), it will be seen that all the arrows point towards the joint. This indicates that the members **b**2, 21 and **a**1 are all in compression because the arrows pointing into the joints are trying to keep the joints apart. The process can be repeated for other joints and the arrows plotted on the truss, to indicate whether the member is in tension. If the arrows point inwards, the member is in tension, and if they point outward, the member is in compression.

Table 4.1 gives a complete analysis of the forces in the truss illustrated in Fig. 4.8. The next stage is to determine the size and shape of the structural members. The tension members will simply depend upon the tensile force and the strength of the material used, but the compression members are more complex as there may be a tendency for the members to buckle before the full strength of the material is reached. Such members will have to be designed as struts, which are dealt with in the next section.

4.3 STRUTS AND TIES

A tie means a member which will always be in tension, and need only be strong enough to resist the 'pull'. A tie member does not have to be stiff, and can be, if necessary, a flexible rod or cable. This can be demonstrated with a piece of string. If the string is pushed, it has no strength at all, but if it is pulled, it shows considerable strength.

On the other hand, a strut is a member which is always in compression, and as such, is liable to buckle. Such a member requires stiffness, so that in a truss, struts can be recognised by their fatness and ties by their slenderness. Therefore:

$$\text{The strength of a tie} = \frac{\text{Force}}{\text{Cross-sectional area}}$$

The strength of a strut = Buckling capacity of strut

The buckling capacity of a strut will be dependent on the slenderness of the strut and the strength of the material. The calculations will be the same as for columns which are dealt with in Chapter 5, and it is not proposed to explain them here.

Worked Example 4.1

Project: Domestic roof truss

Calculation of forces in members

Ref.	Calculations	Output

Question:

Find the forces in a domestic roof truss with a span of 6m, a pitch of 30° and spaced at 1.8 m centres.

App'x 2 and 3

Imposed load = 0.75 kN/m²
Dead load = 1.0 kN/m² } 1.75 kN/m²

Load at each joint = 1.5m × 1.8m × 1.75 kN/m² = 4.7 kN

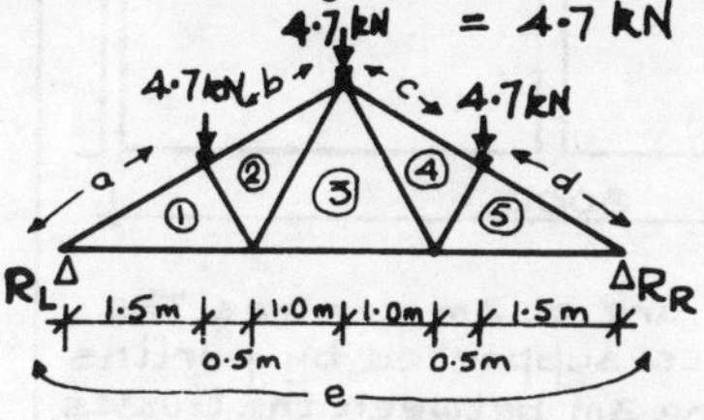

$R_L = R_R = \frac{4.7\text{kN} + 4.7\text{kN} + 4.7\text{kN}}{2} = \underline{7\text{kN}}$

Graphical Solution

Fig. 4.10

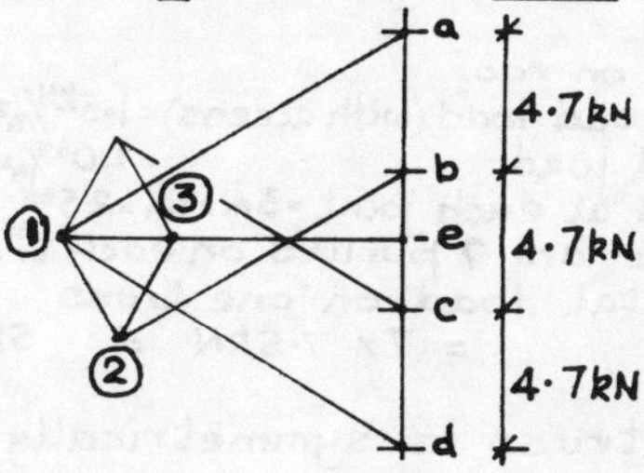

Member	Force	
a 1, d 5	14 kN	Compression
b 1, c 4	11.5 kN	Compression
e 1, e 5	12 kN	Tension
e 3,	8.5 kN	Tension
1 2, 5 4	4 kN	Compression
2 3, 3 4	4 kN	Tension

Worked Example 4.2
Project: Studio/workshop
Forces in structural members of roof truss

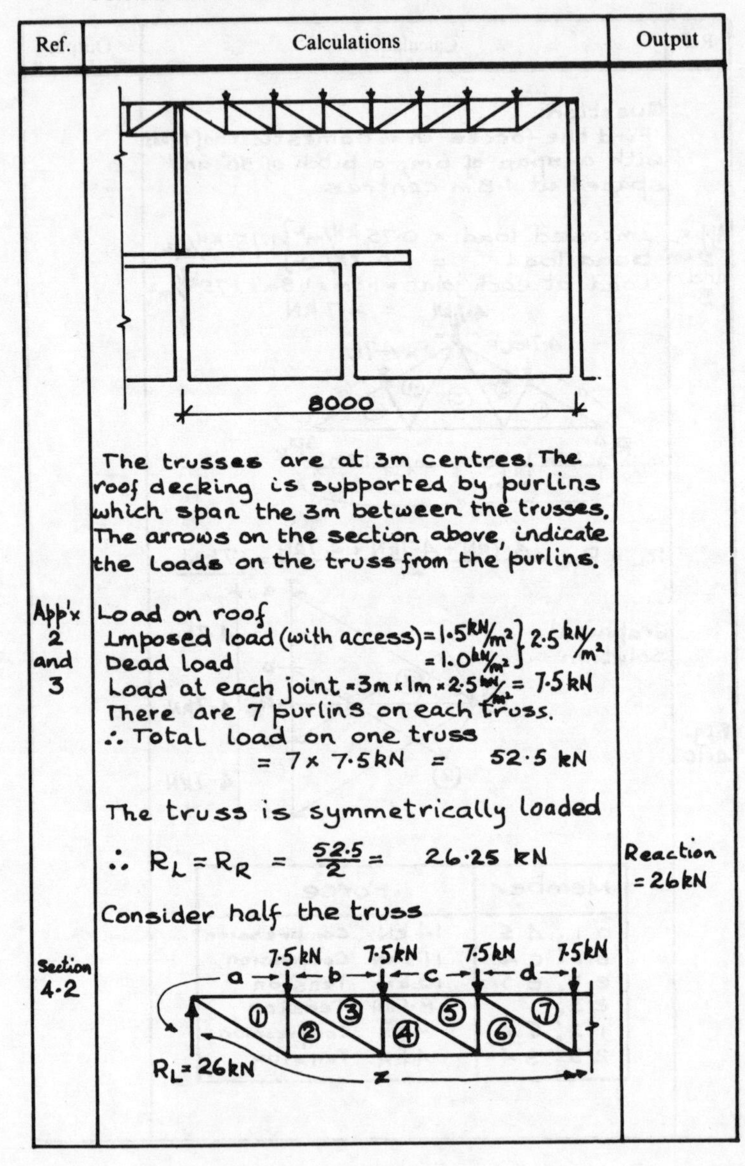

Ref.	Calculations	Output
	8000	
	The trusses are at 3m centres. The roof decking is supported by purlins which span the 3m between the trusses. The arrows on the section above, indicate the loads on the truss from the purlins.	
App'x 2 and 3	Load on roof Imposed load (with access) = $1.5\,kN/m^2$ } $2.5\,kN/m^2$ Dead load = $1.0\,kN/m^2$ Load at each joint = $3m \times 1m \times 2.5\,kN/m^2 = 7.5\,kN$ There are 7 purlins on each truss. $\therefore$ Total load on one truss $= 7 \times 7.5\,kN = 52.5\,kN$	
	The truss is symmetrically loaded $\therefore R_L = R_R = \frac{52.5}{2} = 26.25\,kN$	Reaction = 26kN
Section 4.2	Consider half the truss 7.5kN 7.5kN 7.5kN 7.5kN a b c d 1 2 3 4 5 6 7 $R_L = 26kN$ x	

Worked Example 4.2 continued

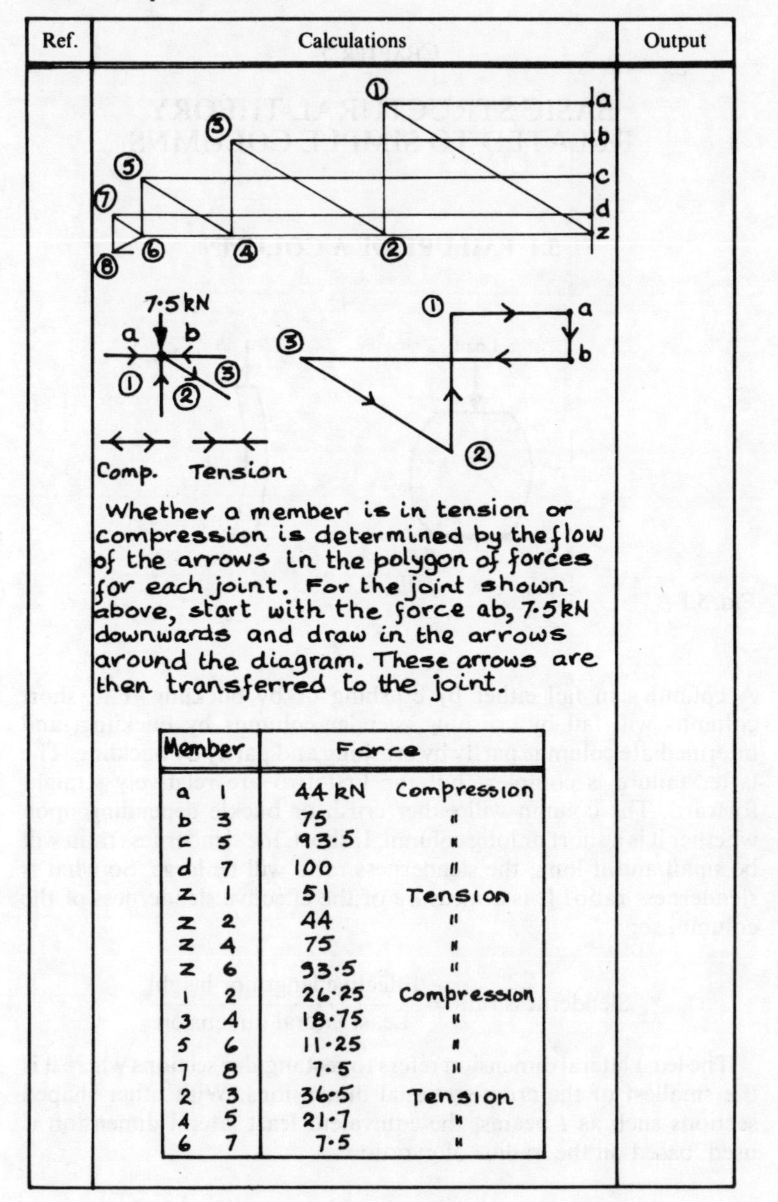

Ref.	Calculations	Output
	Whether a member is in tension or compression is determined by the flow of the arrows in the polygon of forces for each joint. For the joint shown above, start with the force ab, 7·5 kN downwards and draw in the arrows around the diagram. These arrows are then transferred to the joint.	

Member		Force	
a	1	44 kN	Compression
b	3	75	"
c	5	93·5	"
d	7	100	"
z	1	51	Tension
z	2	44	"
z	4	75	"
z	6	93·5	"
1	2	26·25	Compression
3	4	18·75	"
5	6	11·25	"
7	8	7·5	"
2	3	36·5	Tension
4	5	21·7	"
6	7	7·5	"

CHAPTER 5

BASIC STRUCTURAL THEORY RELATED TO SIMPLE COLUMNS

5.1 FAILURE OF A COLUMN

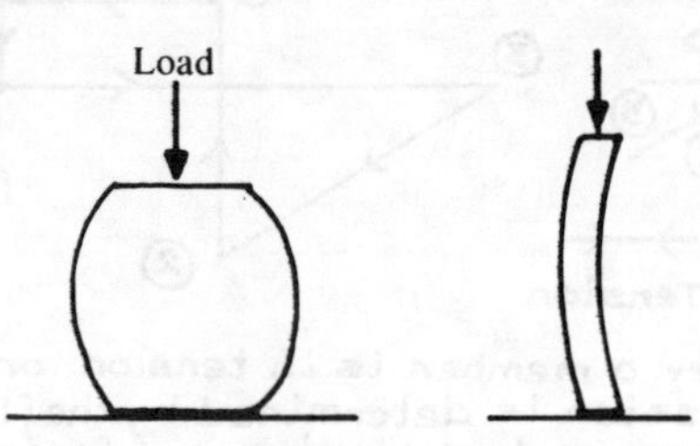

FIG. 5.1

A column can fail either by crushing or by buckling. Fat, short columns will fail by crushing, slender columns by buckling, and intermediate columns partly by crushing and partly by buckling. The latter failure is complex, but the first two are relatively straight forward. The column will either crush or buckle depending upon whether it is a short or long column. If short, the slenderness ratio will be small and if long, the slenderness ratio will be large. So what is slenderness ratio? It is a measure of the effective slenderness of the column, so:

$$\text{Slenderness ratio} = \frac{\text{Effective length or height}}{\text{Least lateral dimension}}$$

The least lateral dimension refers to rectangular sections where it is the smallest of the cross-sectional dimensions. With other shaped sections such as *I* beams, the equivalent least lateral dimension is used, based on the 'radius of gyration'.

$$(\text{Radius of gyration } r_{yy})^2 = \frac{\text{Moment of inertia } I}{\text{Cross-sectional area} A}$$

$$r_{yy} = \left(\frac{I}{A}\right)^{1/2}$$

For example, a concrete column is considered to be a short column if the slenderness ratio is less than 10, and for a brick column, if the slenderness ratio is less than 8. It is assumed that such columns will fail in direct compression. The stress in a short column will be:

$$\text{Actual stress} = \frac{\text{Load}}{\text{Cross-sectional area of column}}$$

This stress then has to be compared with the permissible stress for the materials used, and will be equal to the crushing strength divided by a safety factor, as referred to in section 2.4.

If the column is a 'long column', then it will fail by buckling. The easiest method for calculating the required size for a long column (and this also applies to an intermediate column) is by reducing the permissible basic stress by using reduction factors based on the slenderness ratio of the column (Fig. 5.2).

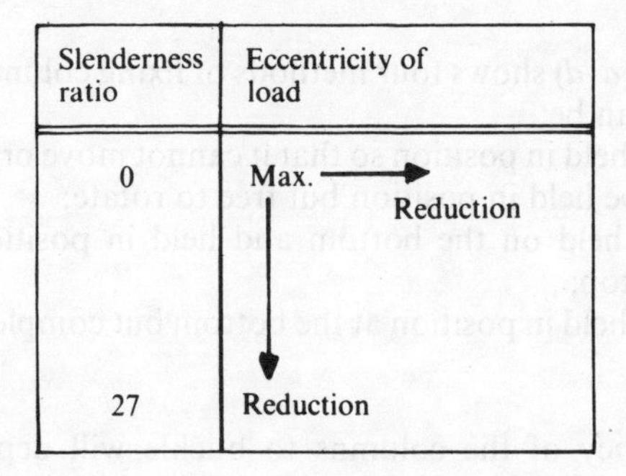

FIG. 5.2

The tendency to buckle will also depend upon whether the column is axially loaded or not. Axially loaded means that the load is placed in the centre. If the load is placed to one side, it will encourage buckling to occur at a much lower stress, and the reduction factor used will have to be that much lower (Fig. 5.2).

5.2 EFFECTIVE LENGTH

It has been seen in section 5.1 how the strength of a compression member will depend upon the slenderness ratio, which itself depends upon the effective length or height. The effective length is a measure of the length of a column that will buckle under load, and this effective length will depend upon how the column is fixed at the top and bottom.

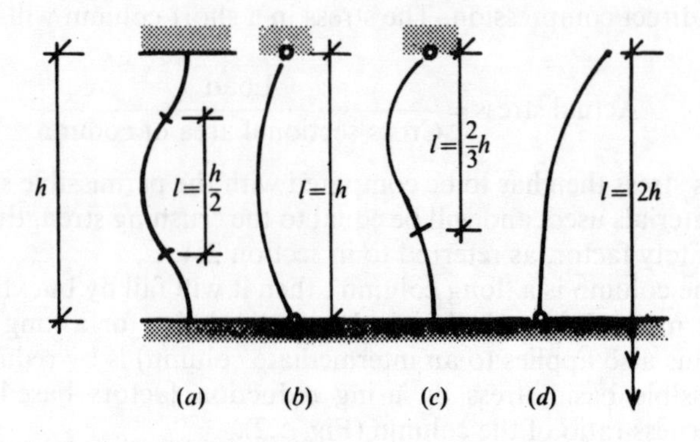

FIG. 5.3. Theoretical effective lengths

Figure 5.3 (*a–d*) shows four methods of fixing columns. The end of the column can be:

(*a*) rigidly held in position so that it cannot move or cannot rotate;

(*b*) it can be held in position but free to rotate;

(*c*) rigidly held on the bottom and held in position but free to rotate at the top;

(*d*) rigidly held in position at the bottom but completely free at the top.

The tendency of the columns to buckle will depend upon the distance between points of 'contra flexure', which is the distance l in Fig. 5.3. The effective lengths given in Fig. 5.3 are the theoretical values and in practice, joints do not always have 100 per cent rigidity. When calculating the effective height of timber, steel, brickwork and concrete columns or compression members in general, the values will be slightly different depending upon which material is being used, but the overall concept will be based on Fig. 5.3.

CHAPTER 6

STRUCTURAL TIMBER DESIGN

There are a variety of softwoods listed in various codes and regulations which are suitable for structural use in buildings. They behave elastically, are good in bending, tension and compression, but some of the softwoods may not be easily purchased, or if they can be, are not up to the required quality. This leaves the structural designer in a quandary. The easiest way out is for the designer to specify a strength rather than specifying a species and grade. If this method is adopted, it is essential that where joists, planks and framing are being used, the builder uses timber, the strength of which can be easily recognised. Therefore, it is strongly recommended that structurally graded timber is used, which can be identified by the marks on each piece of timber. To keep the cost down, it is also recommended to use species which are readily available from timber merchants and this may differ regionally throughout the country.

6.1 STRESS GRADING OF SOFTWOODS

There are two methods used for grading structural timber in Britain, which are set out in reference 6.5 at the end of the chapter. They are:

(*a*) Visual inspection by a trained observer, who considers the number of knots, the size and number of fissures and the general quality of the timber, before giving it a grade, either SS or GS.

SS – special structural grade
GS – general structural grade

(*b*) Machine grading by a machine which measures both the strength and stiffness of a piece of timber without causing any structural damage. There are four machine grades, M75, MSS, M50 and MGS, which may be indicated on the timber by the use of coloured dyes.

M75 – red (75% free of defects)
MSS – purple (special structural)
M50 – blue (50% free of defects)
MGS – green (general structural)

With machine stress graded timber, there is an option for the company carrying out the machine grading to mark each piece of timber with a Strength Class in addition to the species and grade. That is to SC1, SC2, SC3, SC4 and SC5.

Other countries such as Sweden, Finland, Poland and Canada, have their own grading systems, but the general grading principles are the same, with the top grades being suitable for joists and planks while the lower grades are suitable for light framing and studs. The grading system for some of these countries are recognised by the Building Regulations in Britain.

6.2 MOISTURE CONTENT

Many trees when they are first felled have a moisture content well over 150 per cent. In this condition, the timber is described as being green, which is why the permissible stresses are referred to as 'wet' stresses. To reduce this moisture content, the timber is left to season. The natural way to season timber is to allow it to dry out slowly by making sure there is sufficient air to circulate freely round it. The timber increases in strength as it dries out until its moisture content is balanced by the moisture content in the air. This normally is around 18 per cent, and the permissible stresses at this moisture content are referred to as 'dry stresses'.

The seasoning process of timber is rapidly increased by the use of drying kilns, which dry the timber down to a moisture content of below 18 per cent, and at the end of this process, it is often treated with preservatives. The permissible stresses used for structural design calculations, assume that the moisture content is 18 per cent or less.

6.3 DIMENSIONS OF STRUCTURAL TIMBER

The sizes of timber, as specified by the trade, are the sawn timber sizes regardless of whether the timber has been planed, so that the actual size of a planed piece of timber will be somewhat smaller than its specified basic sawn size. Generally, most structural timber is used in the sawn condition, so the basic sizes will be the design sizes. However, if planed or regularised timber is used, as it may be if the structural timber is left exposed, then the reduced cross-sectional dimensions have to be used in the design calculations. Regularised timber is timber which has been resawn or planed to a uniform width. The sectional properties of sawn and planed timber are given in Tables 6.4 and 6.7 on pp. 73 and 76.

There is sometimes some confusion within the trade when architects and engineers refer to the dimensions of timber joists and studs as the breadth and depth. In the trade, cross-sectional sizes are referred to as thickness and width.

When structural design calculations are made, it is important to use sizes of stress-graded timber which are readily available, so that the actual timber sizes the builder uses correspond to the timber sizes assumed in the calculations. These basic sizes are given in Table 6.1.

TABLE 6.1 Basic sawn sizes (for stress-graded timber)

Thickness (mm)	*Width (mm)*						
	75	*100*	*125*	*150*	*175*	*200*	*225*
38	X	X	X	X	X	X	X
44	X	X	X	X	X	X	X
47	X	X	X	X	X	X	X
50	X	X	X	X	X	X	X
63		X	X	X	X	X	X
75		X	X	X	X	X	X

After Table 1, BS 4471.

The other dimension which has to be considered when designing timber structures, is the length of the timber members. Standard lengths of timber are in 300 mm increments with a minimum length of about 1800 mm and a maximum length of 6000 mm although it is difficult to purchase lengths exceeding 5 m. If lengths over 6 m are required, then special orders can be made, but probably the timber will have to be Douglas fir which comes in longer lengths.

6.4 STRENGTH CLASSES

As has been mentioned above, calculations for structural timber design are based on strength requirements, so as to enable the builder to have a reasonably free choice as to where and what timber he purchases. For example, the strength requirement may be met by a high quality piece of spruce or a low quality piece of Douglas fir, both of which may be on offer at the local respectable timber merchant. The builder can then make an economic judgement which one to purchase, and providing the timber has been stress graded and marked, the designer can check the strength class, the species and grade stresses by consulting Tables 6.4 and 6.5 on pp. 73 and 74.

Table 6.5 on p. 74 gives the dry grade stresses and moduli of elasticity for five different strength classes which cover the strength range of softwoods. It will be noticed that the strength of timber will depend on whether the stress is being applied parallel to the grain or perpendicular to it. The grain acts like long, strong fibres which are separated by a softer material. If the fibres are compressed at their ends, they are strong, but if the fibres are compressed together, the softer material in between will tend to fail at a much lower stress. This is why the permissible compressive stresses in the two directions are different.

Table 6.4 (p. 73) gives the strength classes for a selected number of graded species which are most commonly available. As a rough guide, strength classes 1 and 2 are suitable for light framing and studs, while classes 3 and 4 are suitable for joists and planks.

6.5 BEAM DESIGN

Lateral stability of individual beams or joists

Most timber beams are held in position at their ends and held firmly in line along the compression edge by floorboards or other decking materials. Such beams have sufficient lateral support provided the depth to breadth ratio does not exceed 5. If a timber beam is not held in line along its length, then it will not have the same lateral stability and the depth to breadth ratio has to be reduced to 3.

Duration of loading

The strength of timber is affected by the duration of the load. It is much stronger when supporting short-term loads such as wind than it is when supporting long-term loads, such as dead and normal live loads. In a sense, timber has latent strength to deal with temporary loads, which makes it an ideal building material. It is manifested by the use of modification factors, which allow the grade stress in the timber to be increased for short-term loads (Table 6.2, p. 72).

Modification factors

In addition to duration of load, there are a number of other modification factors, most of which apply to 'special timber beams'. For example, if the designer is using straight or curved laminated members, then references 6.1 and 6.2 have to be consulted.

Effective span

The span of flexural members should be taken as the distance between the centres of bearings.

Flexural strength in beams

Actual stress $= f = \frac{M}{Z}$ – see section 3.4

where M = maximum bending moment
Z = section modulus

Permissible stress = Modification factor × Grade stress

The grade stress is obtained from Table 6.5 and can be defined as the stress which can safely be permanently sustained by material of a specific section size and of a particular strength class.

Example

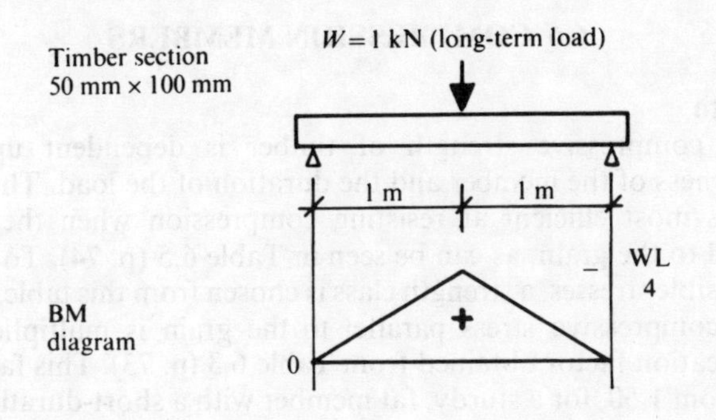

FIG. 6.1. Timber beam

$$\text{Max. bending moment} = WL/4 = \frac{1000\,\text{N} \times 2000\,\text{mm}}{4}$$

$$= 0.5 \times 10^6\,\text{N·mm}$$

$$\text{Section modulus} \quad = Z = \frac{bd^2}{6} \text{ (from Appendix 5)}$$

$$= \frac{50 \times 100 \times 100}{6}$$

$$= 83.3 \times 10^3\,\text{mm}^3$$ (also see Table 6.6 (p. 75))

$$\text{Actual stress} \quad = \frac{M}{Z} = \frac{0.5 \times 10^6}{83.3 \times 10^3} = 6\,\text{N/mm}^2$$

From Table 6.5 (p. 74)

Strenth class SC4 gives Grade stress = 7.5 N/mm^2

Permissible stress = Modification factor × Grade stress
$= 1.0 \times 7.5\,\text{N/mm}^2$
$= 7.5\,\text{N/mm}^2$
$\therefore$ SC4 OK

Therefore from Table 6.4 (p. 73)

use European whitewood to SS grade
or any of the other appropriate species from Table 6.4.

6.6 COMPRESSION MEMBERS

Strength

The compressive strength of timber is dependent upon the slenderness of the member and the duration of the load. The timber itself is most efficient at resisting compression when the load is parallel to the grain, as can be seen in Table 6.5 (p. 74). To find the permissible stresses, a strength class is chosen from this table, and the grade compressive stress parallel to the grain is multiplied by a modification factor obtained from Table 6.3 (p. 73). This factor can vary from 1.50, for a sturdy, fat member with a short-duration load, to 0.10 for a very slender member with a long-term load. If the compression member is held in position at each end, but is free to rotate, then the effective length will be the same as the actual length. The effective thickness will be the minimum width of the member, so that:

$$\text{Slenderness ratio} = \frac{\text{Effective length}}{\text{Minimum width of member}}$$ (see sections 4.3 and 5.2)

Permissible stress = Modification factor × grade stress

$$\text{Actual stress} = \frac{\text{Load}}{\text{Cross-sectional area of member}}$$

Example

$$\text{Actual stress} = \frac{\text{Load}}{\text{Area}} = \frac{20\,000\,\text{N}}{100\,\text{mm} \times 100\,\text{mm}} = 2\,\text{N/mm}^2$$

$$\text{Slenderness ratio} = \frac{\text{Effective length}}{\text{Effective thickness}} = \frac{3000\,\text{mm}}{100\,\text{mm}} = 30$$

From Table 6.3 (p. 73), Modification factor = 0.36

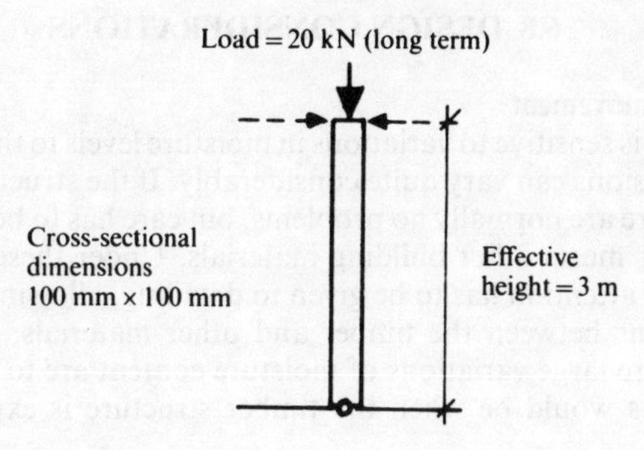

FIG. 6.2. Timber column

From Table 6.5 (p. 74), choose strength class SC3 with grade, compression stress parallel to grain = 6.8 N/mm²

Permissible stress = Modification factor × Grade stress
= 0.36 × 6.8 N/mm²
= 2.4 N/mm²

Permissible stress is greater than the actual stress. Therefore OK.

6.7 FIRE RESISTANCE OF SOLID TIMBER MEMBERS

During a fire, timber suffers a charring effect which produces a very high heat insulation over the unprotected surfaces. In a domestic building, with a 30-minute fire requirement, the depth of the charring will be about 20 mm. If the timber member is large enough, then only the outside of the timber will be affected, leaving a structural core in the middle. The ultimate strength of the timber member during a fire will, therefore, depend upon the stresses in the protected core.

For detailed information, reference 6.3 should be consulted, but in general for domestic buildings the requirements are:

Size of reduced member (most softwoods)

Reduce the dimensions by 20 mm for each exposed surface.

Modification factor

Increase the grade stresses by 2.25 if the width of the member is 70 mm or more.

Deflection

$$\text{Maximum permissible deflection} = \frac{1}{30} \times \text{span}.$$

6.8 DESIGN CONSIDERATIONS

Moisture movement

Timber is sensitive to variations in moisture levels to the extent that the dimensions can vary quite considerably. If the structure is free to move, there are normally no problems, but care has to be taken when the timber meets other building materials. Under these conditions, particular attention has to be given to detailing, allowing for flexible connections between the timber and other materials, especially in areas where large variations of moisture content are to be expected. Such areas would be when the timber structure is exposed to the weather.

Stiffness and deflection

Flexural members should not deflect to the extent of causing damage to surfacing material, or deflect to the extent of causing alarm to the general public. Generally, this will not happen if the deflection is limited to 0.003 of the span, and for purposes of calculating the deflection, the minimum value of the modulus of elasticity should be used for beams acting alone and the mean value used for joists where the load is 'shared'.

For details of deflection equations for various loading conditions, see Appendix 4.

TABLE 6.2 Duration of load

Modification factor	*Duration of load*
1.00	Dead + Imposed
1.25	Dead + snow + Imposed
1.5	Dead + snow + wind + Imposed
1.75	Very short term

After BS 5268: Part 2.

TABLE 6.3 Modification factor for compression members

Slenderness ratio		Modification factors (for SC3 dry timber)	
l/r	*l/b*	*Long term loads*	*Medium term loads*
0–5	1.4	1.0	1.00
5	1.4	0.97	0.97
10	2.9	0.95	0.95
20	5.8	0.90	0.90
30	8.7	0.85	0.84
40	11.5	0.79	0.77
50	14.4	0.72	0.70
60	17.3	0.65	0.61
70	20.2	0.58	0.52
80	23.0	0.51	0.44
90	26.0	0.44	0.38
100	28.8	0.38	0.32
120	34.6	0.29	0.24
140	40.4	0.22	0.18
160	46.2	0.18	0.14

After Table 20 BS 5268 Part 2.

Long term:

$$\frac{\text{E}}{\text{stress}} = \frac{5800}{6.8} = 850$$

Medium term:

$$\frac{\text{E}}{\text{stress}} = \frac{5800}{6.8 \times 1.25} = 680$$

TABLE 6.4 Strength classes and grades

Species	Strength class				
	SC1	*SC2*	*SC3*	*SC4*	*SC5*
Imported					
Douglas fir (Canada)	—	—	GS	SS	—
Hem-fir (Canada)	—	—	GS/M50	SS	M75
Redwood (Europe)	—	—	GS/M50	SS	M75
Whitewood (Europe)	—	—	GS/M50	SS	M75
Spruce-pine-fir (Canada)	—	—	GS/M50	SS/75	
British grown					
Douglas fir	—	GS	M50/SS	—	M75
Scots pine	—	—	GS/M50	SS	M75
Spruce	GS	M50/SS	M75	—	—

After BS 5268: Part 2.

TABLE 6.5 Grade stresses and moduli of elasticity for strength classes: for the dry exposure condition

	Stress parallel to grain				Stress perpendicular to grain	Modulus of elasticity	
Class	Bending (N/mm^2)	Tension (N/mm^2)	Compression (N/mm^2)	Shear (N/mm^2)	Compression (N/mm^2)	Mean (N/mm^2)	Minimum (N/mm^2)
SC1	2.8	2.2	3.5	0.46	1.2	6800	4500
SC2	4.1	2.5	5.3	0.66	1.6	8000	5000
SC3	5.3	3.2	6.8	0.67	1.7	8800	5800
SC4	7.5	4.5	7.9	0.71	1.9	9900	6600
SC5	10.0	6.0	8.7	1.00	2.4	10700	7100

After BS 5268: Part 2.

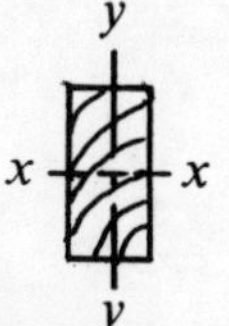

TABLE 6.6 Sectional properties of sawn timber

Basic sizes (mm)	Area (mm^2)	Second moment of area I_{xx} ($10^6 mm^4$)	Second moment of area I_{yy} ($10^6 mm^4$)	Section modulus Z_{xx} ($10^3 mm^3$)
38 × 75	2850	1.3	0.34	35.6
38 × 100	3800	3.2	0.46	63.3
38 × 125	4750	6.2	0.57	99.0
38 × 150	5700	10.7	0.68	142
38 × 175	6650	17.0	0.80	194
38 × 200	7600	25.3	0.91	254
38 × 225	8550	36.1	1.03	321
44 × 75	3300	1.5	0.53	41.2
44 × 100	4400	3.7	0.71	72.3
44 × 150	6600	12.4	1.06	165
44 × 200	8800	29.3	1.42	293
47 × 75	3520	1.6	0.65	44.0
47 × 100	4700	3.9	0.86	78.3
47 × 125	5870	7.6	1.08	122
47 × 150	7050	13.2	1.30	176
47 × 175	8220	21.0	1.51	240
47 × 200	9400	31.3	1.73	313
47 × 225	10570	44.6	1.94	396
50 × 75	3750	1.7	0.78	46.9
50 × 100	5000	4.2	1.04	83.3
50 × 125	6250	8.1	1.30	130
50 × 150	7500	14.1	1.56	188
50 × 175	8750	22.3	1.82	255
50 × 200	10000	33.3	2.08	333
50 × 225	11200	47.5	2.34	422
63 × 150	9450	17.7	3.13	236
63 × 175	11000	28.1	3.65	322
63 × 200	12600	42.0	4.17	420
63 × 225	14200	59.8	4.69	532
75 × 150	11200	21.1	5.27	281
75 × 175	13100	33.5	6.15	383
75 × 200	15000	50.0	7.03	500
75 × 225	16900	71.2	7.91	633

TABLE 6.7 Sectional properties of planed timber

Basic sawn sizes (mm)	*Minimum planed sizes (mm)*	*Area (mm^2)*	*Second moment of area* I_{xx} *($10^6 mm^4$)*	*Second moment of area* I_{yy} *($10^6 mm^4$)*	*Section modulus* Z_{xx} *($10^3 mm^3$)*
38 × 75	35 × 72	2520	1.1	0.25	30.2
38 × 100	35 × 97	3400	2.6	0.34	54.9
38 × 125	35 × 120	4200	5.0	0.43	84.0
38 × 150	35 × 145	5080	8.9	0.52	123
38 × 175	35 × 169	5920	14.1	0.60	167
38 × 200	35 × 194	6790	21.3	0.69	220
38 × 225	35 × 219	7660	30.6	0.78	280
44 × 75	41 × 72	2950	1.3	0.41	35.4
44 × 100	41 × 97	3980	3.1	0.55	64.3
44 × 150	41 × 145	5940	10.4	0.83	144
44 × 200	41 × 194	7950	24.9	1.11	257
47 × 75	44 × 72	3168	1.3	0.51	38.0
47 × 100	44 × 97	4268	3.3	0.69	69.0
47 × 125	44 × 120	5280	6.3	0.85	105
47 × 150	44 × 145	6380	11.2	1.03	154
47 × 175	44 × 169	7436	17.7	1.20	209
47 × 200	44 × 194	8536	26.7	1.37	276
47 × 225	44 × 219	9636	38.5	1.55	351
50 × 75	47 × 72	3380	1.4	0.62	40.6
50 × 100	47 × 97	4560	3.5	0.84	73.7
50 × 125	47 × 120	5640	6.7	1.04	113
50 × 150	47 × 145	6820	11.9	1.25	165
50 × 175	47 × 169	7940	18.9	1.46	224
50 × 200	47 × 194	9120	28.6	1.68	295
50 × 225	47 × 219	10300	41.1	1.89	376
63 × 150	60 × 145	8700	15.2	2.61	210
63 × 175	60 × 169	10100	24.1	3.04	286
63 × 200	60 × 194	11600	36.5	3.49	376
63 × 225	60 × 219	13100	52.5	3.94	480
75 × 150	72 × 145	10400	18.3	4.51	252
75 × 175	72 × 169	12200	29.0	5.26	343
75 × 200	72 × 192	14000	43.8	6.03	452
75 × 225	72 × 219	15800	63.0	6.81	576

Worked Example 6.1
Project: Studio house
Calculation of roof beams

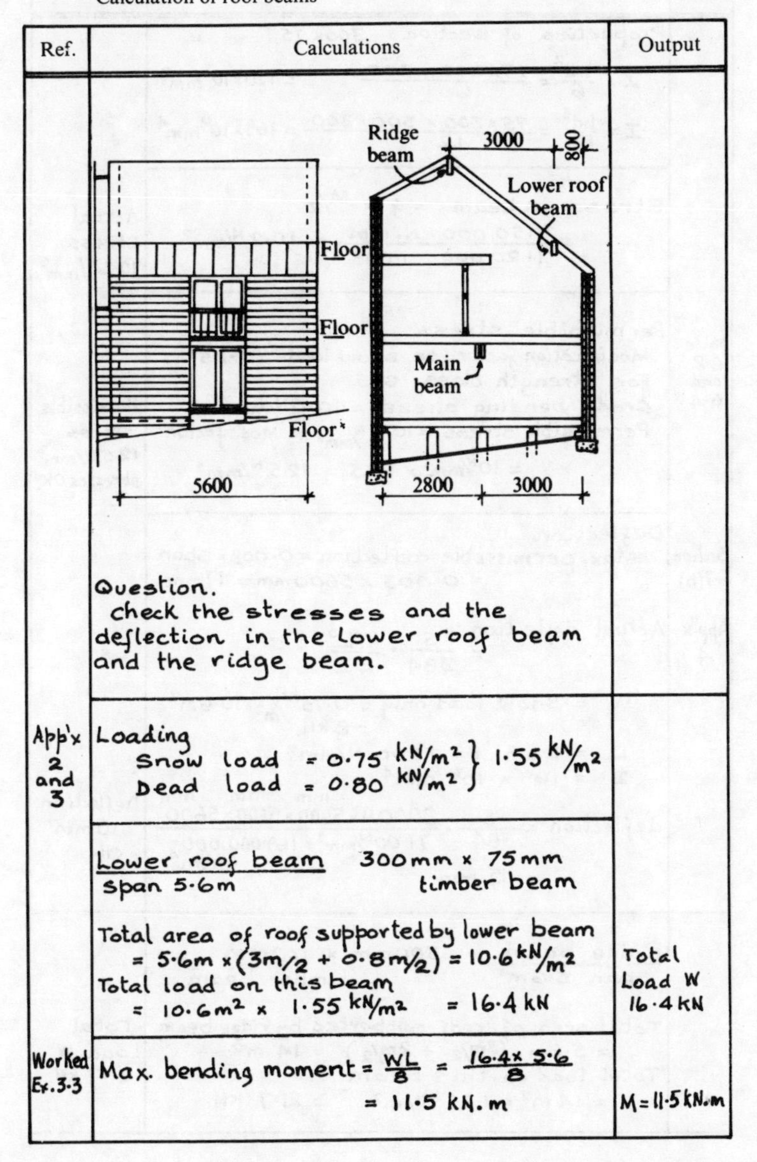

Ref.	Calculations	Output
	Drawing: Ridge beam; 3000; 800; Lower roof beam; Floor; Floor; Main beam; Floor; 5600; 2800; 3000	
	Question. check the stresses and the deflection in the lower roof beam and the ridge beam.	
App'x 2 and 3	Loading Snow load $= 0{\cdot}75$ kN/m² Dead load $= 0{\cdot}80$ kN/m² } $1{\cdot}55$ kN/m²	
	Lower roof beam 300mm × 75mm timber beam span 5·6m	
	Total area of roof supported by lower beam $= 5{\cdot}6\text{m} \times (3\text{m}/2 + 0{\cdot}8\text{m}/2) = 10{\cdot}6\ \text{kN/m}^2$ Total load on this beam $= 10{\cdot}6\text{m}^2 \times 1{\cdot}55\ \text{kN/m}^2 = 16{\cdot}4\ \text{kN}$	Total Load W 16·4 kN
Worked Ex. 3·3	Max. bending moment $= \frac{WL}{8} = \frac{16{\cdot}4 \times 5{\cdot}6}{8}$ $= 11{\cdot}5\ \text{kN.m}$	M = 11·5 kN.m

Worked Example 6.1 continued

Ref.	Calculations	Output
	Properties of section 300× 75 $Z = \frac{bd^2}{6} = \frac{75 \times 300 \times 300}{6} = 1120 \times 10^3\ mm^3$ $I = \frac{bd^3}{12} = \frac{75 \times 300 \times 300 \times 300}{12} = 169 \times 10^6\ mm^4$	
	Stress in beam $= f = M/Z$ $= \frac{1150\,000\ N.mm}{1120\,000\ mm^3} = 10.2\ N/mm^2$	Actual stress $10.2\ N/mm^2$
Table 6.2 and 6.4	Permissible stress Modification factor for snow load = 1·25 For strength class SC5 Grade bending stress $= 10.0\ N/mm^2$ Permissible stress $= 10.0\ N/mm^2 \times$ Mod. factor $= 10\ N/mm^2 \times 1.25 = 12.5\ N/mm^2$	Permissible stress $12.5\ N/mm^2$ stresses OK.
Section 6.7(b) App'x 4	Deflection Max. permissible deflection = 0·003 × span = 0·003 × 5600 mm = 17 mm Actual deflection $= \frac{5}{384} \cdot \frac{WL^3}{EI}$ W = Snow load only $= 0.75\ kN/m^2 \times 10.6\ m^2$ = 8 kN. L = 5·6 m $E = 7100\ N/mm^2$ $I = 169 \times 10^6\ mm^4$ deflection $= \frac{5}{384} \cdot \frac{8000\,N \times 5600\,mm \times 5600\,mm \times 5600\,mm}{7100\,N/mm^2 \times 169\,000\,000\,mm^4}$ = 15 mm	Deflection 15 mm OK.
	Ridge beam 300 mm × 100 mm Span 5·6 m timber beam Total area of roof supported by ridge beam $= 5.6\,m \times (3\,m/2 + 2\,m/2) = 14\ m^2$ Total load on this beam $= 14\ m^2 \times 1.55\ kN/m^2 = 21.7\ kN$	Total Load W 21·7 kN.

Worked Example 6.1 continued

Ref.	Calculations	Output
	Max. bending moment $= \frac{WL}{8} = \frac{21{\cdot}7 \times 5{\cdot}6}{8}$ $= 15{\cdot}2$ kN. m	M = 15·2 KN.m
	Properties of section 300 × 100 $Z = \frac{bd^2}{6} = \frac{100 \times 300 \times 300}{6} = 1500 \times 10^3\ mm^3$ $I = \frac{bd^3}{12} = \frac{100 \times 300 \times 300 \times 300}{12} = 225 \times 10^6\ mm^4$	
	Stress in beam $= f = M/Z$ $= \frac{15200\,000\ N.mm}{1500\,000\ mm^3} = 10{\cdot}1\ N/mm^2$	Actual stress 10·1 N/mm²
from above	Permissible stress $= 12{\cdot}5\ N/mm^2$	stress OK.
from above	Deflection Max. permissible deflection = 17 mm Actual deflection $= \frac{5}{384} \cdot \frac{WL^3}{EI}$ W = snow load only $= 0{\cdot}75\ kN/m^2 \times 14\ m^2$ $= 10{\cdot}5$ kN. L = 5·6m E = 7100 N/mm² $I = 252 \times 10^6\ mm^4$ deflection $= \frac{5}{384} \cdot \frac{10500\,N \times 5600\,mm \times 5600\,mm \times 5600\,mm}{7100\,N/mm^2 \times 225\,000\,000\,mm^4}$ = 15 mm. deflection OK.	Deflection 15 mm

Worked Example 6.2
Project: Terrace house
Sizes of timber floor joists

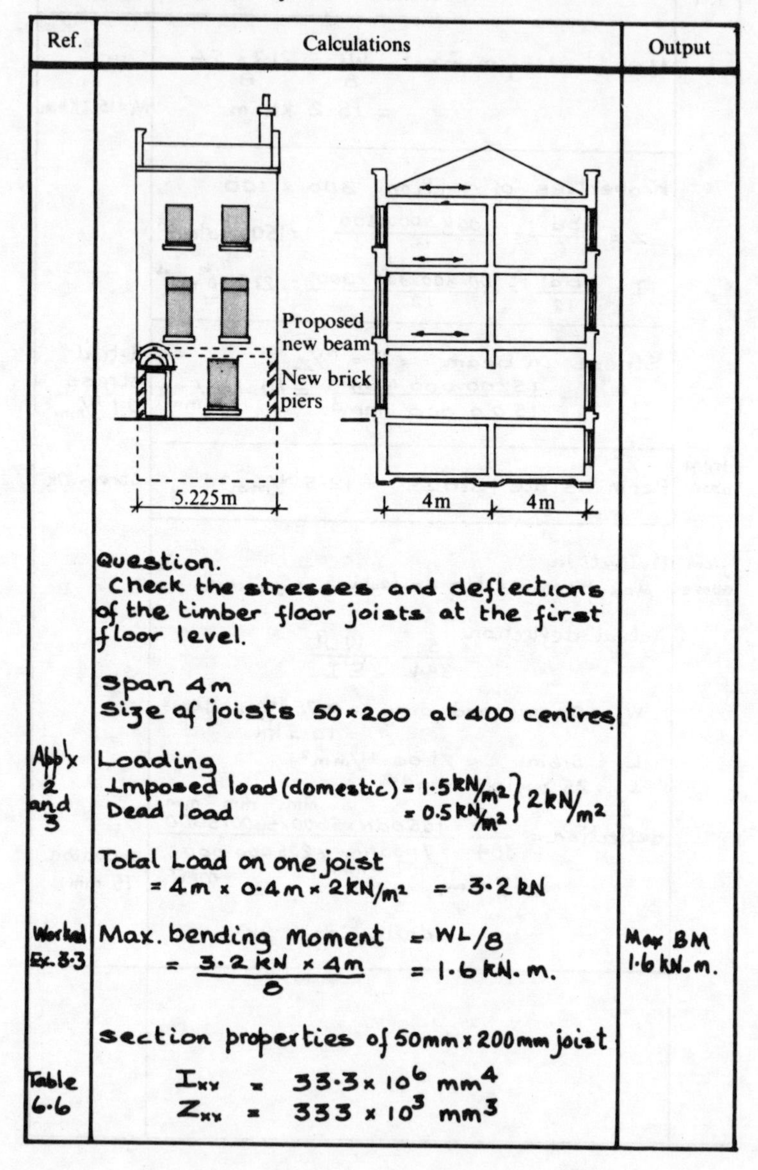

Ref.	Calculations	Output
	Question. Check the stresses and deflections of the timber floor joists at the first floor level.	
	Span 4m Size of joists 50×200 at 400 centres	
App'x 2 and 3	Loading Imposed load (domestic) = 1.5 kN/m² } 2 kN/m² Dead load = 0.5 kN/m²	
	Total Load on one joist = 4m × 0.4m × 2kN/m² = 3.2 kN	
Worked Ex. 8.3	Max. bending moment = WL/8 = $\frac{3.2\ kN \times 4m}{8}$ = 1.6 kN.m.	Max BM 1.6 kN.m.
	section properties of 50mm × 200mm joist	
Table 6.6	I_{xx} = 33.3 × 10⁶ mm⁴ Z_{xx} = 333 × 10³ mm³	

Worked Example 6.2 continued

Ref.	Calculations	Output
	Actual stress $= f = M/Z$ $= \dfrac{1600\,000 \text{ N.mm}}{333\,000 \text{ mm}^3} = 4.8 \text{ N/mm}^2$	
Tables 6.4 and 6.5	If joists are European Whitewood and grade GS or M50, the strength class SC3, gives a grade bending stress $= 5.3 \text{ N/mm}^2$	∴ stress OK
Section 6.7(b) App'x 4	Deflection Max. permissible deflection $= 0.003 \times$ span $= 0.003 \times 4000 \text{mm} = 12 \text{ mm}$ Actual deflection $= \dfrac{5}{384} \cdot \dfrac{W L^3}{E I}$ $W = 3.2 \text{ kN}$ $L = 4.0 \text{ m}$ $E = 8800 \text{ N/mm}^2$ $I = 33.3 \times 10^6 \text{ mm}^4$ Deflection $= \dfrac{5}{384} \cdot \dfrac{3200 \text{ N} \times 4000 \text{ mm} \times 4000 \text{ mm} \times 4000 \text{ mm}}{8800 \text{ N/mm}^2 \times 33300\,000 \text{ mm}^4}$ $= 9 \text{ mm}$ ∴ Deflection OK.	Deflection 6mm OK.
	Note: The grade stress could be increased by a modification factor of 1.1 as a load sharing system.	

Worked Example 6.3
Project: Cottage extension
Sizes of floor joists

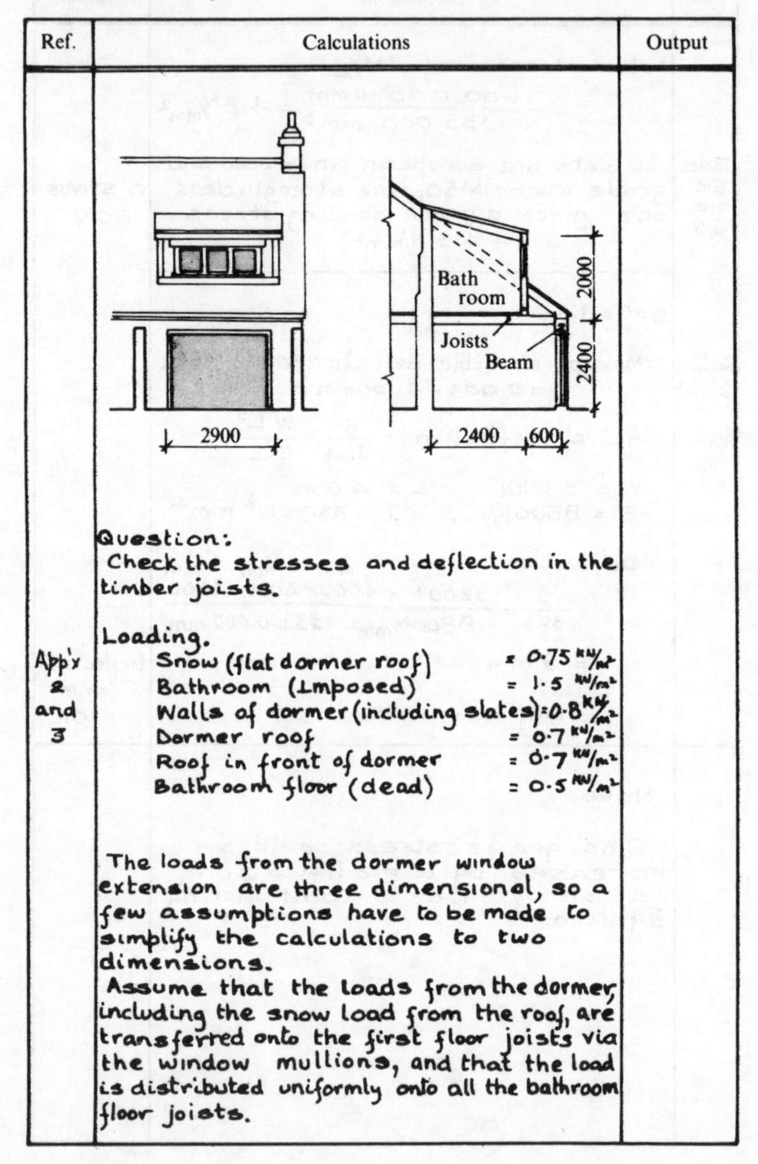

Ref.	Calculations	Output
	Question: Check the stresses and deflection in the timber joists.	
App'x 2 and 3	**Loading.** Snow (flat dormer roof) = 0·75 kN/m² Bathroom (Imposed) = 1·5 kN/m² Walls of dormer (including slates) = 0·8 kN/m² Dormer roof = 0·7 kN/m² Roof in front of dormer = 0·7 kN/m² Bathroom floor (dead) = 0·5 kN/m²	
	The loads from the dormer window extension are three dimensional, so a few assumptions have to be made to simplify the calculations to two dimensions. Assume that the loads from the dormer, including the snow load from the roof, are transferred onto the first floor joists via the window mullions, and that the load is distributed uniformly onto all the bathroom floor joists.	

Worked Example 6.3 continued

Ref.	Calculations	Output
	Proposed bathroom floor joists size of joists 50 × 175 at 400 centres	
	Assume a 400 mm width of bathroom floor and dormer roof including roof in front of dormer.	
	Area of 400mm wide strip of floor or roof = 0·4m × 3m = 1·2m²	
	Area of 400mm vertical strip of dormer wall/window = 0·4m × 2m = 0·8m²	
	Total Load of roof:	
from above	Dead = 1·2m² × 0·7 kN/m² = 0·84 Snow = 1·2m² × 0·7 kN/m² = 0·9 } 1.74 kN	
	Half this load only is supported on dormer window = 0·87 kN.	
	Total Load of floor:	
from above	Dead = 1·2m² × 0·5 kN/m² = 0·6 Imposed = 1·2m² × 1·5 kN/m² = 1·8 } 2·4 kN	
	Total Load of dormer wall/window Dead = 0·8m² × 0·8 kN/m² = 0·64 kN	
	Roof: 0·87 kN Wall/window: 0·64 kN Floor: 2·4 kN or 0·8 kN/m run R_L — 2·4 m — 0·6 m — R_R	
	Reactions. Take moments about R_R 3m × R_L = 0·6 × 1·5 kN + 1·5m × 2·4 kN = 4·5 kN·m $R_L = \frac{4.5\text{ kN·m}}{3\text{m}}$ = 1·5 kN	R_L = 1·5 kN
	R_R = Total Load − R_L = 2·4 kN	R_R = 2·4 kN

Worked Example 6.3 continued

Ref.	Calculations	Output
Section 3.3	1·5 kN; 0·8 kN per m run; $R_L = 1·5$ kN; $R_R = 2·4$ kN; 2·4 m; 0·6 m S.F. diagram: 1·5 +; 1·875 m; −; 2·4 BM diagram: 0; M_{max} $M_{max} = 1·875 \times R_L - 0·8 \times 1·875 \times \frac{1·875}{2}$ $= 2·8 - 0·9 = 1·9$ kN.m	Max. B.M = 1·9 kN.m
Table 6.6	Joists 50 × 175 $Z_{xx} = 255 \times 10^3$ mm³ $I_{xx} = 22·3 \times 10^6$ mm⁴ Actual stress $= f = \frac{M}{Z}$ $= \frac{1900\ 000 \text{ N.mm}}{255\ 000 \text{ mm}^4} = 7.45$ N/mm²	
Table 6.5 and 6.4 Table 6.2	Permissible stress Assume strength class SC4 such as Hem-fir (Canada) SS grade Then grade bending stress = 7·5 N/mm² Modification factor for duration of load = 1·25 Permissible stress = 1·25 × 7·5 = 9·4 N/mm² Stress OK.	Use 50×175 SS grade Joists.

Worked Example 6.3 continued

Ref.	Calculations	Output
	Deflection.	
Section 6.7(b)	Max. permissible deflection = 0.003 × span = 0.003 × 3000 mm = 9 mm	permissible deflection = 9mm
	Actual deflection	
App'x 4	a) For a uniform load defl. $= \frac{5}{384} \cdot \frac{WL^3}{EI}$	
	b) For a point load defl. $= \frac{PL^3}{48EI}\left\{\frac{3a}{L} - 4\left(\frac{a}{L}\right)^3\right\}$	
	a) W = 2.4 kN L = 3 m E = 9900 N/mm² I = 22.3 × 10⁶ mm⁴	
	defl. $= \frac{5}{384} \cdot \frac{2400\,N \times 3000\,mm \times 3000\,mm \times 3000\,mm}{9900\,N/mm^2 \times 22300\,000\,mm^4}$ = 3.8 mm	
	b) P = 1.5 kN L = 3 m E = 9900 N/mm² I = 22.3 × 10⁶ mm⁴ a = 0.6 m a/L = 0.2	
	$\left\{\frac{3a}{L} - 4\left(\frac{a}{L}\right)^3\right\} = 0.6 - 0.032 = 0.568$	
	defl. $= \frac{1500\,N \times 3000\,mm \times 3000\,mm \times 3000\,mm \times 0.568}{48 \times 9900\,N/mm^2 \times 22300\,000\,mm^4}$ = 2.2 mm	
	Max. deflection = 3.8 + 2.2 = 6 mm deflection OK	deflection = 6mm.

Worked Example 6.4
Project: Shop extension
Timber joists supporting plant

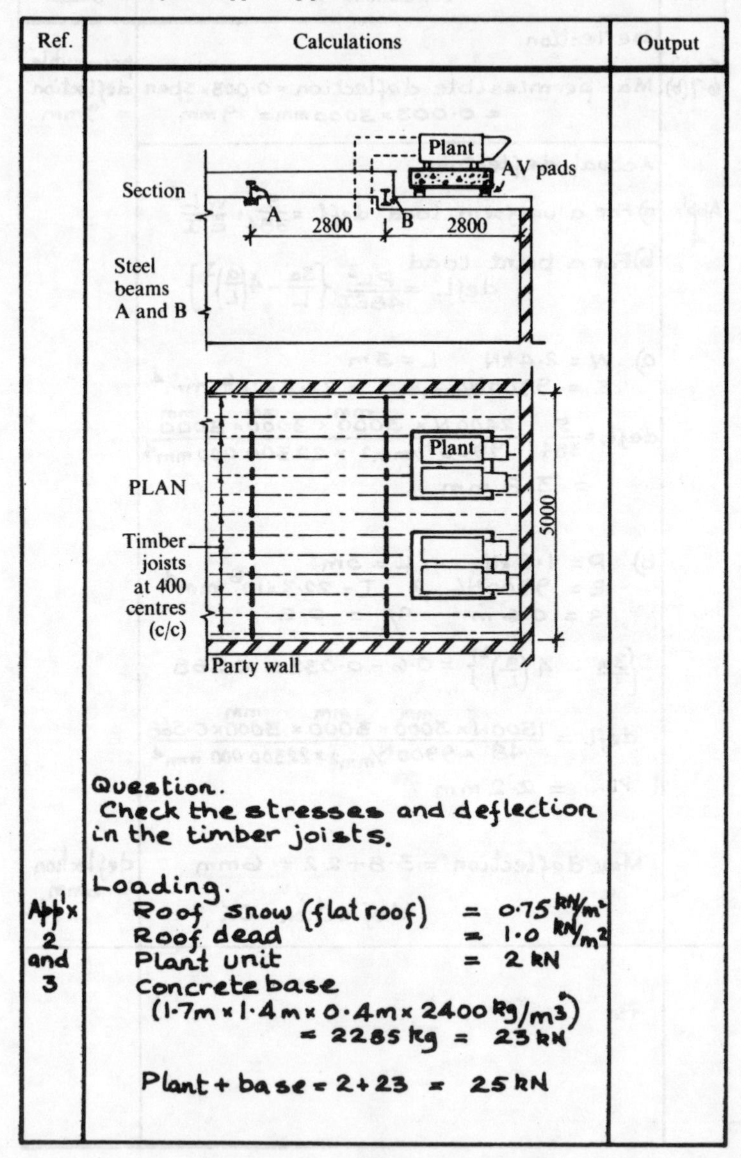

Ref.	Calculations	Output
	Question. Check the stresses and deflection in the timber joists.	
App'x 2 and 3	Loading. Roof snow (flat roof) = 0·75 kN/m² Roof dead = 1·0 kN/m² Plant unit = 2 kN Concrete base (1·7m × 1·4m × 0·4m × 2400 kg/m³) = 2285 kg = 23 kN Plant + base = 2 + 23 = 25 kN	

Worked Example 6.4 continued

Ref.	Calculations	Output
	The load from each plant base is transferred to bearers via the A.V. pads. (anti-vibration). The bearers distribute the load over four joists.	
	Consider the timber joists supporting the plant and concrete base. Size of joists 50 × 175 at 400 centres	
	Area of 'snow' over plant base $= 1.7\text{m} \times 1.4\text{m} = 2.4\ \text{m}^2$	
	Snow load on base $= 2.4 \times 0.75 = 1.8$ kN	
	Load carried by one joist $= 0.45$ kN	
	Plant + base load $= 25$ kN	
	Load carried by one joist $= 6.25$ kN	
	Area of roof carried on one joist $= 0.4\text{m} \times 2.8\text{m} = 1.1\ \text{m}^2$	
	Dead load of roof carried by one joist $= 1.1\text{m}^2 \times 1.0\ \text{kN/m}^2 = 1.1$ kN	
	The snow load at the ends of the joists is ignored.	
	Snow: 0.45 kN (over 1700) Plant and base: 6.25 kN Dead: 1.1 kN R_L 1500 R_R 2800	
	Reactions.	
	Total Load $= 0.45 + 6.25 + 1.1 = 7.8$ kN	$R_L = 3.9$ kN
	$R_L = R_R = 3.9$ kN	$R_R = 3.9$ kN

Worked Example 6.4 continued

Ref.	Calculations	Output
	3·35 kN, 3·35 kN, 1·1 kN, R_L = 3·9 kN, R_R = 3·9 kN B.M. diagram	
App'x 4	Max. BM = for uniform load + for point load $= \frac{WL}{8} + Pa$ W = 1·1 kN L = 2·8 m P = 3·35 kN a = 650 mm $\frac{WL}{8} = \frac{1\cdot1\text{ kN} \times 2\cdot8\text{ m}}{8} = 0\cdot39$ kN.m Pa = 3·9 kN × 0·65 m = 2·5 kN.m Max BM = 0·39 + 2·5 = 2·9 kN.m	
Table 6·6	Joists 75 × 175 $Z_{xx} = 383 \times 10^3$ mm³ $I_{xx} = 33\cdot5 \times 10^6$ mm⁴ Actual stress = $f = \frac{M}{Z}$ $= \frac{2\,900\,000\text{ N.mm}}{383\,000\text{ mm}^3} = 7\cdot6\text{ N/mm}^2$	Actual stress = 7·6 N/mm²
Table 6·5 and 6·4	Permissible stress Assume strength class SC5 such as Douglas fir M75 grade Then grade bending stress = 10 N/mm²	Use 75 × 175 M75 grade joists

Worked Example 6.4 continued

Ref.	Calculations	Output
Section 6·7(b)	Deflection Max. permissible deflection = 0·003 × span = 0·003 × 2800mm = 8 mm	permissible deflection = 8mm
App'x 4	Actual deflection. a) For uniform load $\text{defl.} = \frac{5}{384} \cdot \frac{WL^3}{EI}$ b) For two point loads $\text{deflection} = \frac{PL^3}{6EI}\left\{\frac{3a}{4L} - \left(\frac{a}{L}\right)^3\right\}$	
Table 6·4	a) W = 1·1 kN L = 2·8m $E = 10700\ \text{N/mm}^2$ $I = 33{\cdot}5 \times 10^6\ \text{mm}^4$ $\text{defl.} = \frac{5}{384} \cdot \frac{1100\text{N} \times 2800\text{mm} \times 2800\text{mm} \times 2800\text{mm}}{10700\ \text{N/mm}^2 \times 33500\,000\ \text{mm}^4}$ = 1 mm. b) P = 3·35 kN L = 2·8 m $E = 10700\ \text{N/mm}^2$ $I = 33{\cdot}5 \times 10^6\ \text{mm}^4$ a = 0·65 m a/L = 0·23 $\left\{\frac{3a}{4L} - \left(\frac{a}{L}\right)^3\right\} = 0{\cdot}17 - 0{\cdot}01 = 0{\cdot}16$ $\text{defl.} = \frac{3350\text{N} \times 2800\text{mm} \times 2800\text{mm} \times 2800\text{mm} \times 0{\cdot}16}{6 \times 10700\ \text{N/mm}^2 \times 33500\,000\ \text{mm}^4}$ = 5 mm Max deflection = 1mm + 5mm = 6mm deflection OK	Deflection 6mm

Worked Example 6.5
Project: Roof purlin
Bending stresses, deflection and fire resistance

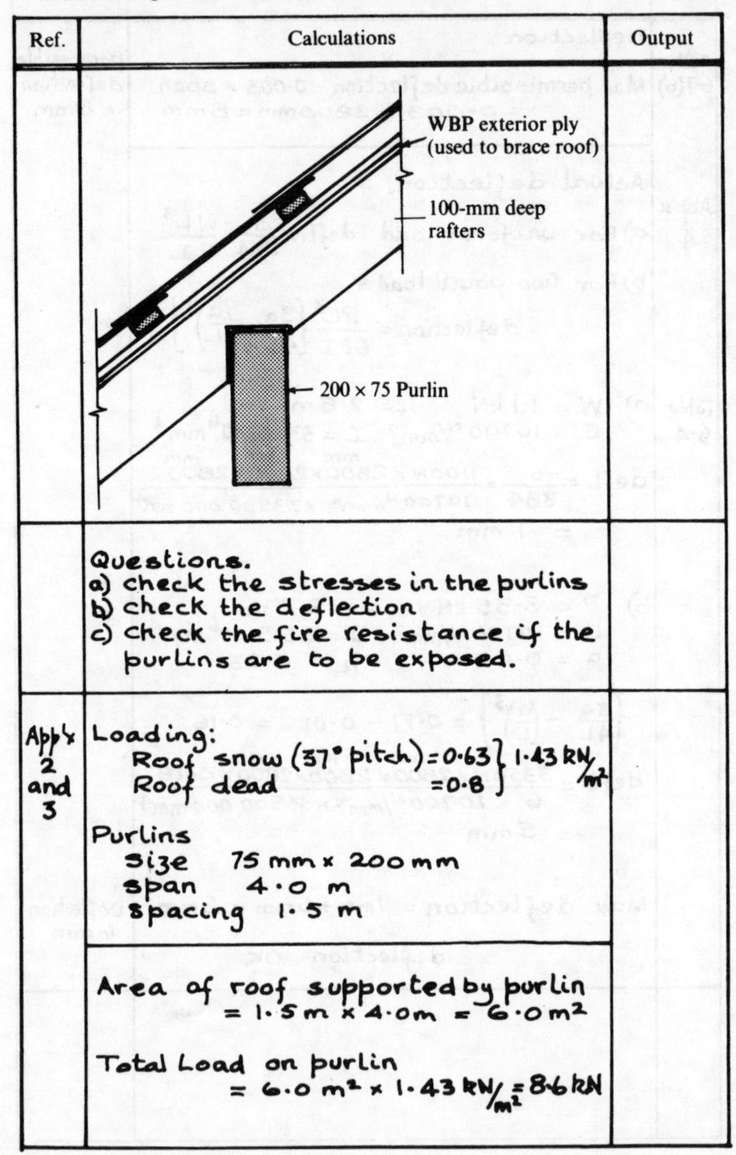

Ref.	Calculations	Output
	WBP exterior ply (used to brace roof) 100-mm deep rafters 200 × 75 Purlin	
	Questions. a) check the stresses in the purlins b) check the deflection c) check the fire resistance if the purlins are to be exposed.	
App'x 2 and 3	Loading: Roof snow (37° pitch) = 0·63 } 1·43 kN/m² Roof dead = 0·8 Purlins Size 75 mm × 200 mm Span 4·0 m Spacing 1·5 m	
	Area of roof supported by purlin = 1·5 m × 4·0 m = 6·0 m² Total Load on purlin = 6·0 m² × 1·43 kN/m² = 8·6 kN	

Worked Example 6.5 continued

Ref.	Calculations	Output
Worked Ex 3.3	Bending moment $M_{max} = \frac{WL}{8} = \frac{8.6\text{kN} \times 4\text{m}}{8} = 4.3\text{kN.m}$	Maximum $M = 4.3$ kN. m.
Table 6.7	Purlin 75 mm × 200 mm As the purlins are to be exposed, the surface will have to be planed. ∴ Use sectional properties of planed timber. $I_{xx} = 43.8 \times 10^6 \text{ mm}^4$ $Z_{xx} = 452 \times 10^3 \text{ mm}^3$ Actual stress $= f = \frac{M}{Z}$ $= \frac{4\,300\,000 \text{ N.mm}}{452\,000 \text{ mm}^3} = 9.5 \text{ N/mm}^2$	Actual Stress 9.5 N/mm^2
Table 6.5 and 6.4 Table 6.2	Permissible stress Assume strength class SC5 such as Scots pine M75 grade Then grade bending stress = 10 N/mm^2 Modification factor for duration of load $= 1.25$ ∴ permissible stress $= 1.25 \times 10.0 \text{ N/mm}^2$ $= 12.5 \text{ N/mm}^2$	permissible Stress 12.5 N/mm^2
Section 6.7(b)	Deflection Max. permissible deflection $= 0.003 \times$ span $= 0.003 \times 4000 = 12 \text{ mm}$	permissible deflection = 12 mm
App'x 4 Table 6.4	Actual deflection $= \frac{5}{384} \cdot \frac{WL^3}{EI}$ Total imposed load on purlin $= W = 6.0 \text{m}^2 \times 0.63 \text{ kN/m}^2 = 3.8 \text{ kN}$ $L = 4 \text{m}$ $E = 10700 \text{ N/mm}^2$ $I = 43.8 \times 10^6 \text{ mm}^4$	N.B. Assume purlins and rafters act as 'shared' system. E = mean

Worked Example 6.5 continued

Ref.	Calculations	Output
	$\text{deflection} = \frac{5}{384} \cdot \frac{3800\text{N} \times 4000\text{mm} \times 4000\text{mm} \times 4000\text{mm}}{10700\text{N/mm}^2 \times 43800000\text{mm}^4} = 7\text{mm}$ It should be noted that the deflection is based on the imposed load only. In this example, the combined loads of dead and imposed will produce a deflection slightly in excess of 12mm. To avoid any visual 'sag', the purlin should be propped during construction.	Deflection = 7mm
	Charring effect of Purlins. Solid timber members build up fire resistance by the charring of the timber surface. The solid timber in the middle remains as a structural member. The purlin in this example is exposed on three sides.	
BS 5268 Part 4 clause 4.2.1	Solid members For 30 min. charring rate:- 20mm off exposed surfaces. Exposed Surfaces 35mm; 200 mm; 180 mm; 75mm Section properties of charred member $Z_{xx} = \frac{bd^2}{6} = \frac{35 \times 180^2}{6} = 189 \times 10^3\ \text{mm}^3$ $I_{xx} = \frac{bd^3}{12} = \frac{35 \times 180^3}{12} = 17 \times 10^6\ \text{mm}^4$	

Worked Example 6.5 continued

Ref.	Calculations	Output
As before	Max. bending moment = 4.3 kN.m Actual stress in charred purlin $= f = \frac{M}{Z} = \frac{4\,300\,000 \text{ N.mm}}{189\,000 \text{ mm}^3}$ $= 22.6 \text{ N/mm}^2$	Charred stress 22.6 N/mm^2
BS 5268 clause 5.1.2 (b) Table 6.2	Permissible stress under charring When width of beam is 70 mm or greater Modification factor for charring = 2.25 Modification factor for 'snow' load = 1.25 ∴ permissible stress under charring = 2.25 × 1.25 × grade stress = 2.25 × 1.25 × 10 N/mm² = 34 N/mm² Fire resistance OK	permissible charred stress 34 N/mm^2
clause 5.1.1 (b) 5.1.2 (c)	Deflection must not exceed $\frac{1}{30}$ × span $= \frac{1}{30} \times 4000 = 133$ mm Actual deflection under charring $= \frac{5}{384} \cdot \frac{WL^3}{EI}$ W = 8.6 kN L = 4 m E = 10700 N/mm² I = 17 × 10⁶ mm⁴ $\text{defl.} = \frac{5}{384} \cdot \frac{8600\text{N} \times 4000\text{mm} \times 4000\text{mm} \times 4000\text{mm}}{10700 \text{ N/mm}^2 \times 17\,000\,000 \text{ mm}^4}$ = 40 mm.	Max permissible charred deflection 133 mm Max. charred deflection 40 mm

REFERENCES

6.1 *Structural Use of Timber*, British Standard 5268: Part 2, 1984. Permissible stress, design, materials and workmanship.

6.2 *The Structural Use of Timber*, British Standard Code of Practice, CP 112: Part 2, 1971.

6.3 *Structural Use of Timber*, British Standard 5268: Part 4, section 4.1, 1978. Method of calculating fire resistance of timber members.

6.4 *Specification for sizes of sawn and processed softwood*, British Standard 4471: 1987.

6.5 *Specification for softwood grades for structural use*, British Standard 4978: 1988.

CHAPTER 7

STRUCTURAL STEEL DESIGN

7.1 STRUCTURAL STEEL

Hot-rolled sections

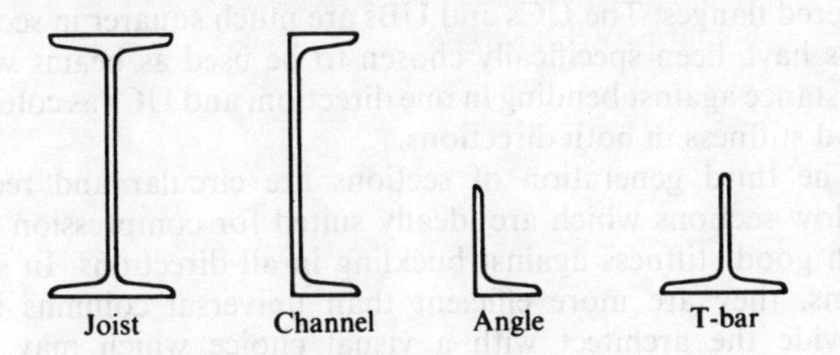

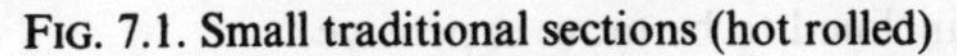

FIG. 7.1. Small traditional sections (hot rolled)

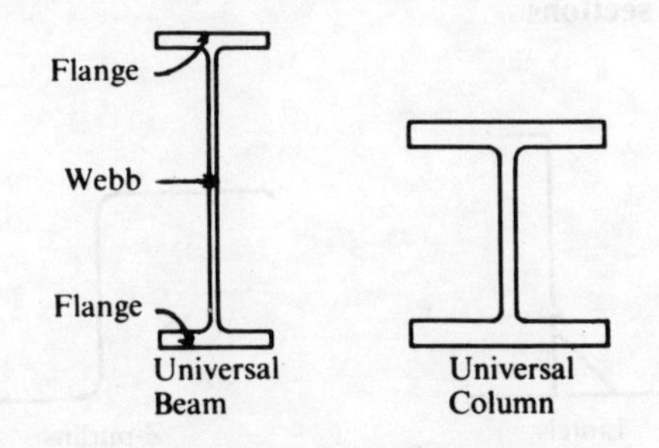

FIG. 7.2. Larger traditional sections (hot rolled)

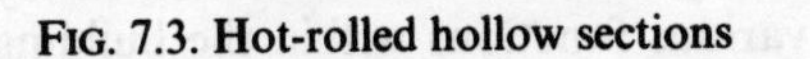

FIG. 7.3. Hot-rolled hollow sections

Figures 7.1, 7.2 and 7.3 show three generations of hot-rolled sections, the rolled steel joist (RSJ), the universal column (UC) and beam (UB), and finally the hollow steel sections. Up to the mid 1950s, the steel sections were based on the RSJ, and these were used both for beams and for columns, but as such were not totally suited for either. Something new was needed.

When the second generation of sections were introduced in the form of UBs and UCs, the larger RSJs were discontinued, but the smaller sections were kept, although their name was changed to joists. Therefore, joists, channels, angles and T-bars are all based on the first-generation sections, and their shapes is distinguished by their tapered flanges. The UCs and UBs are much squarer in section. The UBs have been specifically chosen to be used as beams with good resistance against bending in one direction, and UCs as columns with good stiffness in both directions.

The third generation of sections are circular and rectangular hollow sections which are ideally suited for compression members with good stiffness against buckling in all directions. In structural terms, they are more efficient than universal columns and they provide the architect with a visual choice which may be more pleasing, but they are inclined to be more expensive.

Cold-formed sections

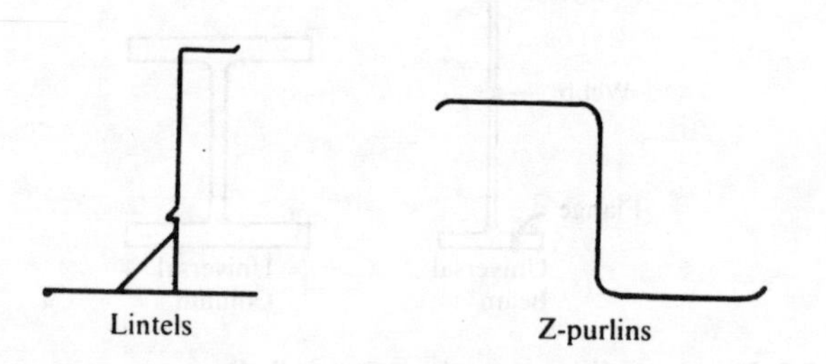

FIG. 7.4. Cold-formed sections

Cold-formed sections have been developed from the car industry, and have become acceptable in the building industry mainly through the development of modern protective coatings, which are able to protect the sections from corrosion. A few examples of cold-formed sections are shown in Fig. 7.4, the most common are Z-purlins and standard lintels. There are many other sections fabricated by numerous manufacturers for various functions within the building industry.

Grades of steel

There are three main grades of steel commonly used in the manufacture of structural steel sections in Britain. They are grade 43, otherwise known as mild steel, and grades 50 and 55 which are known as high-strength steels. For the size of building projects illustrated in this chapter, it is normal to use mild steel.

Manufacture of steel sections

After steel has been produced to one of the above grades and with the appropriate properties for structural steel, it is cast as steel ingots. The ingots are reheated and rolled into 10 m long solid 'blooms', which can be reheated and rolled into sheet steel suitable for cold-worked sections or rolled directly into traditional steel sections. The hollow sections are made from strip or plate steel by being first cold worked into a circular form and then welded down the seam. The section is then reheated and rolled into the required shapes.

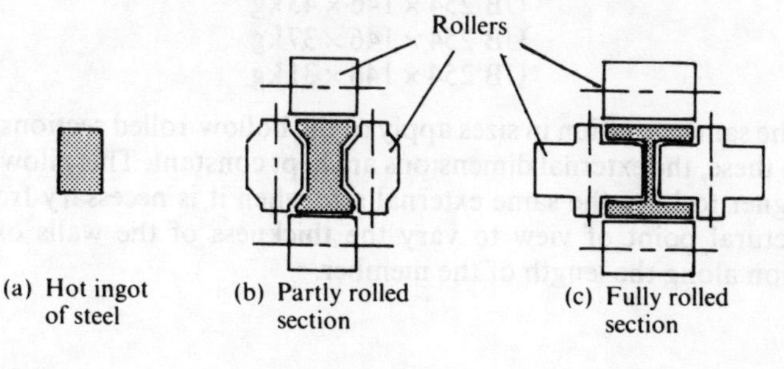

FIG. 7.5

It is worth considering a little further how the hot-rolled sections are manufactured to shed some light on the variations on the shapes and sizes listed in the structural steel handbooks. Consider the working of the rollers shown in Fig. 7.5. The white-hot rectangular steel bloom (a), starts on its journey along the rollers. The first rollers (b), begin to push the bloom into shape and at the same time the bloom starts to stretch and becomes longer. The rollers continue to push the steel into shape until the required cross-section is achieved (c), which meets the dimensional specifications of the required steel section. The useful point to understand is that the same set of rollers can produce a number of cross-sectional variations with only small adjustments to the production line. For example, the thickness of the top and bottom flanges can be altered by lifting and lowering the top and bottom rollers, and if one of the side rollers is replaced by a straight roller, a channel section can be produced (Fig. 7.6).

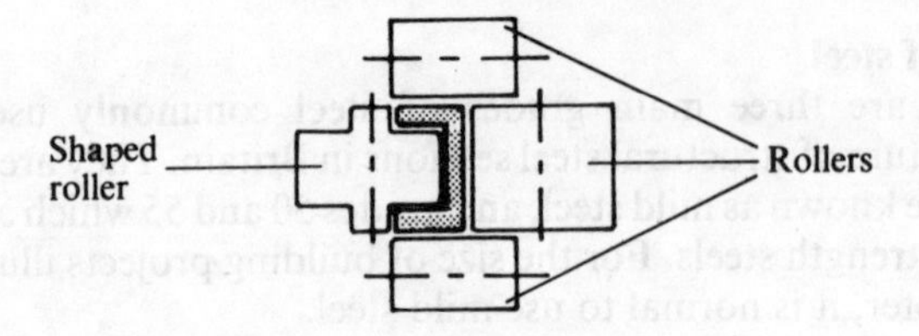

FIG. 7.6. Fully rolled channel section

The specified sizes of structural steel sections are based on nominal or serial sizes, and within this size there may be a range of weights as a consequence of different thicknesses of flanges and webs. It is therefore, normal to refer to a steel section by its nominal or serial size and by its weight per metre length to identify it from similar sizes but different weights. For example, a serial size of 254 mm by 146 mm will reveal three different UBs, all with a different weight per metre length.

UB 254 × 146 × 43 kg
UB 254 × 146 × 37 kg
UB 254 × 146 × 31 kg

The same variation in sizes apply to the hollow-rolled sections, but with these, the external dimensions are kept constant. This allows the designer to keep the same external size when it is necessary from a structural point of view to vary the thickness of the walls of the section along the length of the member.

7.2 BEAM SIZES USING SAFE-LOAD TABLES

A convenient way to determine the sizes of beam required for a particular project, is to use safe-load tables such as the tables set out in the BCSA and SCI handbook. However, care must be exercised when using these tables. For example, it must be remembered that these tables are for uniformly loaded, simply supported beams having full lateral restraint. If the beam is not laterally restrained, then they can only be used as a guide. What is meant by lateral restraint is discussed in section 7.4.

Before these tables are used, it is first necessary to calculate the total load on the beam, by using the loading information from Appendices 2 and 3. It is emphasised that it is the total uniformly distributed load acting on the beam which is required. Once this has been calculated, a beam is selected from the tables with a 'safe load' greater than the calculated value, so that the beam selected is 'safe'.

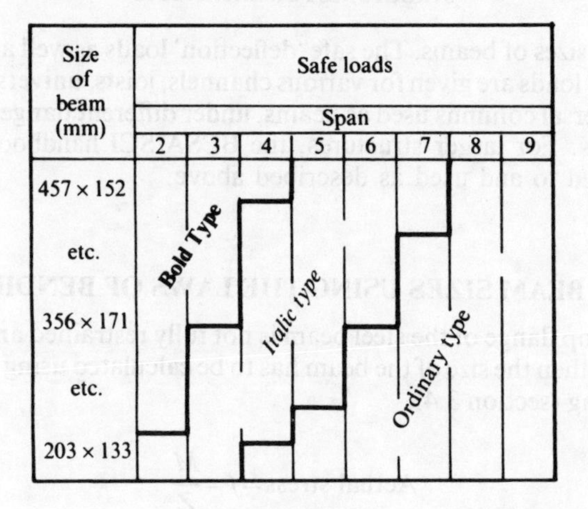

FIG. 7.7

Figure 7.7 is a sketch of the safe-load tables. The sizes of beam are down the left-hand column, the spans are along the top and the safe loads are tabulated in the middle. Also along the top are deflection coefficients which will be required if the deflections have to be checked as described in section 7.5.

Having checked the span of the beam from the drawing, select the nearest span listed along the top of the safe-load tables. Then look down the column below this span until the tabulated safe load only slightly greater than the calculated load is found, and locate the section size in the left-hand column corresponding to this safe load. If the safe load is printed in *italic* type this will be the size of beam required and there will be no necessity to perform further checks. However, if the safe load is in **bold** or ordinary type, further calculations will have to be made. The bold type means that the web of the beam will buckle or crush under the safe load, so web stiffening would be required. If this is the case, the designer would be advised to consult a structural engineer or use a larger section.

If the safe load is printed in ordinary type, this means that excessive deflection will occur under the safe load, causing finishes, such as plaster to crack. So ordinary type means check the deflections. To prevent cracking, the deflection has to be kept to within 1/360 of the span, and if the load W_D which causes this amount of deflection is calculated, it can be compared with the actual total uniformly distributed load on the beam. W_D is the safe load based on the deflection.

The safe loads for a selection of beam sizes have been calculated and tabulated in Table 7.4 (p. 108). This table has been prepared for domestic sizes of beams. The safe 'deflection' loads as well as the safe 'strength' loads are given for various channels, joists, universal beams and universal columns used as beams, under different flange restraint conditions. For larger structures, the BCSA/SCI handbook has to be referred to and used as described above.

7.3 BEAM SIZES USING THE LAWS OF BENDING

If the top flange of the steel beam is not fully restrained and held in position, then the size of the beam has to be calculated using the Laws of Bending (section 3.4):

$$\text{Actual stress} = f = \frac{M}{Z}$$

where M = max. bending moment
Z = elastic modulus

The actual maximum stress in the beam is compared with the permissible stresses for mild steel as given in Table 7.5 (p. 114), and if the actual stress is less, then the beam is safe. The permissible stress will depend upon the longitudinal slenderness of the beam (l/r_y) and the effective vertical stiffness of the flange (D/T), where

l = effective length of the compression flange
r_y = radius of gyration (effective width of the I-beam) about an axis lying in the plane of bending
D = overall depth of beam
T = mean thickness of flange

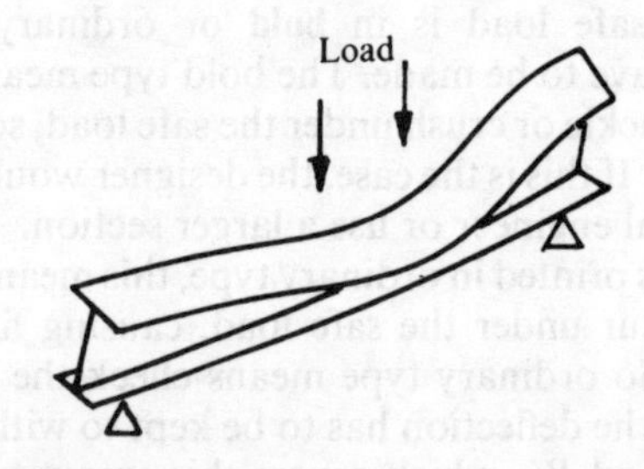

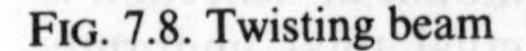
FIG. 7.8. Twisting beam

The slenderness is a reflection of the stiffness, and the more slender the compression flange is, the weaker the beam will be. This is demonstrated in Fig. 7.8. When the load is increased, the top flange begins to twist and buckle causing the beam to fail, even though the stress in the beam at the point of failure may be considerably lower than if the top flange was restrained from buckling. Therefore, if the top flange is unrestrained, the permissible stresses will be reduced to those values given in Table 7.5 (p. 114), if the section is made from mild steel. For other grades of steel, consult reference 7.3.

The values of r_y and D/T are obtained from the section tables, such as Tables 7.2. and 7.3 (pp. 105 and 106). The effective length (l) will depend upon how the beam is supported. For beams simply supported and restrained at each end against torsion, which they will be if the ends are built into a wall, then:

$$\text{Effective length } l = \text{Span}$$

And if there is no torsional restraint at the ends, then:

Effective length to be increased by 20 per cent.

7.4 EFFECTIVE LATERAL SUPPORT OF COMPRESSION FLANGE

Concrete slabs and steel beams

When steel beams support reinforced concrete slabs, the friction between the slab and the beams is sufficient to give the compression flanges full lateral support.

Timber joists and steel beams

When steel beams support timber floors with the joists sitting on the beams, the friction is not sufficient to give the compression flange full lateral restraint. Even if the timber joists are notched around the steel beam, there is always the possibility of the timber shrinking as it dries out. For timber joists to give full lateral support, there must be a proper connection such as the joists being bolted to the beam or being very well wedged and notched.

Encased in concrete

If the steel beam is encased in concrete, the lateral stability of the top compression flange will be improved. The concrete and reinforcement around the beam helps to stiffen the beam and reduce its slenderness ratio. If the beam is correctly encased and reinforced, (clause 21 in reference 7.3), so that there is at least 50 mm cover to the steel beam, then the radius of gyration r_y may be taken as $0.2(b + 100\,\text{mm})$. This will greatly decrease the slenderness ratio and hence increase the permissible compressive stress in bending (Table 7.5, p. 114). However, the stress should not exceed one and a half times that permitted for the uncased section.

It is, therefore, important to remember when using the BCSA and SCI safe-load tables, that they do assume the top compression flange of the beam to be fully restrained. If the flange is not restrained, the safe-load tables, as has been mentioned before, are only a guide, and the actual safe load has to be calculated.

7.5 DEFLECTION

Maximum beam deflection must be limited to prevent damage to finishes and to avoid the beam looking unsafe. If the effects of the deflection due to the dead load can be nullified by pre-cambering the beam or by straightening the external casing, then the deflection limits will depend upon the imposed loads only. If the dead-load deflections cannot be nullified, then it is advisable to apply the deflection limits to both the dead and imposed loads.

To prevent damage to finishes, the deflection of a steel beam should be limited to 1/360 of the span. The load which produces this deflection is called the deflection load W_D, and will be the safe load the beam can carry based on the deflection limits.

$$\text{The deflection load } W_D = \frac{CI}{1000}$$

where C = deflection coefficient
I = moment of inertia measured in cm^4 and obtained from the section property Tables 7.1, 7.2 and 7.3 (pp. 103, 105 and 106)

This equation is derived from the basic deflection formula for a uniformly loaded simply supported beam given in Appendix 4. That is:

$$\text{Deflection} = 5/384 \cdot \frac{WL^3}{EI}$$

where W = total uniformly distributed load, kN
L = span in metres
E = Young's modulus (210000 N/mm^2 for mild steel)
i = moment of inertia for steel section, cm^4
I = moment of inertia

$$\text{If deflection} = \frac{\text{span}}{360}$$

$$W = \frac{384\,EI}{5 \times 360\,L^2} = \frac{384\,E}{5 \times 0.36\,L^2}\,\frac{I}{1000} = \frac{CI}{1000}$$

where C = deflection coefficient for each span

TABLE 7.1

Span L (m)	1.5	2.0	2.5	3.0	3.5	4.0	4.5	5.0	5.5	6.0
Coefficient C	199	112	71.7	49.8	36.6	28.0	22.1	17.9	14.8	12.4

7.6 SHEAR

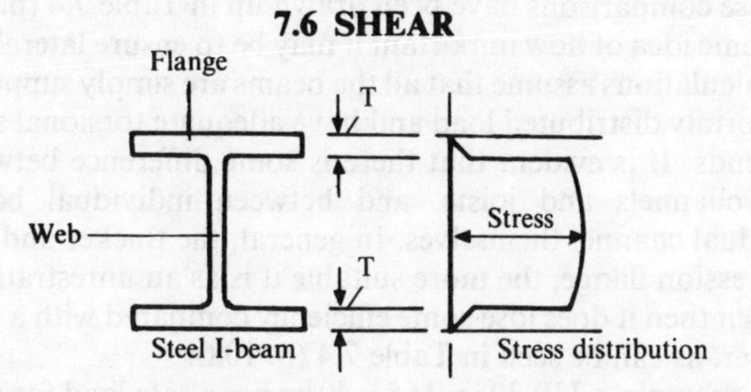

FIG. 7.9. Distribution of shear stress

The distribution of shear stress in an I-beam is shown in Fig. 7.9. It will be observed that nearly all the shear stress is carried by the web so it can be assumed that:

$$\text{Average shear stress} = \frac{\text{Shear force}}{\text{Area of web}}$$

This calculated average shear stress must be less than the allowable average shear stress for mild steel (grade 43) which is 100 N/mm^2.

7.7 COMPARISON OF SAFE LOADS

In this chapter, it has been shown how the maximum load a beam can carry will depend upon a number of design parameters. It is, therefore, appropriate to make a comparison between these maximum 'safe' loads for a selection of joists, channels, UBs and UCs used as beams. A 'guesstimated' economic span for each section has been chosen, based on the rule of thumb method to determine the depth of beams described in section 1.4. For example, a beam lightly loaded spanning 4 m will have a depth equal to 200 mm, based on a span to depth ratio of 20. For a heavy load, the depth will be 225 mm, based on a span to depth ratio of 18.

The parameters considered are:

(*a*) the top compression flange being fully restrained along its entire length;

(*b*) the top compression flange being unrestrained;

(*c*) the maximum deflection being limited to 1/360 of the span;

(*d*) the steel beam being encased in concrete in accordance with clause 21, reference 7.3.

These comparisons have been drawn up in Table 7.4 (p. 108), and give some idea of how important it may be to ensure lateral restraint. The calculations assume that all the beams are simply supported with a uniformly distributed load and have adequate torsional stiffness at their ends. It is evident that there is some difference between UBs, UCs, channels and joists, and between individual beams and individual channel themselves. In general, the thicker and wider the compression flange, the more suitable it is as an unrestrained beam, but even then it does lose some efficiency compared with a restrained member, as can be seen in Table 7.4 (p. 108).

For example, a UB 305 × 165 × 40 kg has a safe load for a 6 m span equal to 123 kN, but this is reduced to 70 kN if the compression flange is unrestrained. Also UBs, whose flanges are only 102 mm wide, give reasonable safe loads when the compression flange is restrained, but very poor safe loads when the top flange is not restrained. From the evidence given in Table 7.4 (p. 108), these narrow UBs should not be used in the unrestrained condition. A beam with a wider flange would be more suitable.

The safe loads based on deflection are for the total uniform load on the beams. If the deflection due to dead loads can be lost by pre-cambering, etc., then these loads become the imposed safe loads, which will mean that the total safe loads due to deflection will be the value tabulated plus the total dead load.

The tabulated safe loads based on the sections being encased in solid concrete, are assumed to have no further bracing. This would be the case, if the beam was supporting a brick wall without any lateral bracing from the floor above. Then the only lateral restraint the compression flange would have, would be from the concrete and reinforcement surrounding the beam.

Generally, UCs are only used as beams when the load is quite large and the available depth is small. The section in Table 7.4 (p. 108), dealing with UCs, includes a number of different spans with span to depth ratios varying from 20 up to 33. For all these UCs, the critical load is the one based on the maximum permitted deflection. It will also be seen that it does not really matter whether the compression flange is restrained or not. This is due to the very wide flanges all UCs have, making it very unlikely for them to fail by buckling.

UNIVERSAL BEAMS

Designation		Depth of section D	Width of section B	Thickness		Area of Section	I Moment of inertia Axis x–x	Z Elastic modulus Axis x–x	r_y Radius of gyration Axis y–y
Serial size	Mass per metre			Web t	Flange T				
(mm)	(kg)	(mm)	(mm)	(mm)	(mm)	(cm^2)	(cm^4)	(cm^3)	(cm)
305 × 165	54	310.9	166.8	7.7	13.7	68.4	11 710	753.3	3.94
	46	307.1	165.7	6.7	11.8	58.9	9948	647.9	3.90
	40	303.8	165.1	6.1	10.2	51.5	8523	561.2	3.85
305 × 127	48	310.4	125.2	8.9	14.0	60.8	9504	612.4	2.75
	42	306.6	124.3	8.0	12.1	53.2	8143	531.2	2.70
	37	303.8	123.5	7.2	10.7	47.5	7162	471.5	2.67
305 × 102	33	312.7	102.4	6.6	10.8	41.8	6487	415.0	2.15
	28	308.9	101.9	6.1	8.9	36.3	5421	351.0	2.08
	25	304.8	101.6	5.8	6.8	31.4	4387	287.9	1.96
254 × 146	43	259.6	147.3	7.3	12.7	55.1	6558	505.3	3.51
	37	256.0	146.4	6.4	10.9	47.5	5556	434.0	3.47
	31	251.5	146.1	6.1	8.6	40.0	4439	353.1	3.35
254 × 102	28	260.4	102.1	6.4	10.0	36.2	4008	307.9	2.22
	25	257.0	101.9	6.1	8.4	32.2	3408	265.2	2.14
	22	254.0	101.6	5.8	6.8	28.4	2867	225.7	2.05
203 × 133	30	206.8	133.8	6.3	9.6	38.0	2887	279.3	3.18
	25	203.2	133.4	5.8	7.8	32.3	2356	231.9	3.10

A selection of UBs from reference 7.1.

TABLE 7.2(b) Section properties: Universal Columns

UNIVERSAL COLUMNS
To: BS4: Part 1

Designation		*Depth of section*	*Width of section*	*Thickness*		*Area of Section*	*I Moment of inertia Axis x–x*	*Z Elastic modulus Axis x–x*	r_y *Radius of gyration Axis y–y*
Serial size	*Mass per metre*	*D*	*B*	*Web t*	*Flange T*				
(mm)	*(kg)*	*(mm)*	*(mm)*	*(mm)*	*(mm)*	*(cm²)*	*(cm⁴)*	*(cm³)*	*(cm)*
254×254	167	289.1	264.5	19.2	31.7	212.4	29914	2070	6.79
	132	276.4	261.0	15.6	25.3	167.7	22575	1634	6.68
	107	266.7	258.3	13.0	20.5	136.6	17510	1313	6.57
	89	260.4	255.9	10.5	17.3	114.0	14307	1099	6.52
	73	254.0	254.0	8.6	14.2	92.9	11360	894.5	6.46
203×203	86	222.3	208.8	13.0	20.5	110.1	9462	851.5	5.32
	71	215.9	206.2	10.3	17.3	91.1	7647	708.4	5.28
	60	209.6	205.2	9.3	14.2	75.8	6088	581.1	5.19
	52	206.2	203.9	8.0	12.5	66.4	5263	510.4	5.16
	46	203.2	203.2	7.3	11.0	58.8	4564	449.2	5.11
152×152	37	161.8	154.4	8.1	11.5	47.4	2218	274.2	3.87
	30	157.5	152.9	6.6	9.4	38.2	1742	221.2	3.82
	23	152.4	152.4	6.1	6.8	29.8	1263	165.7	3.68

A selection of UCs from reference 7.1.

TABLE 7.3(a) Section properties: joists

JOISTS
To BS4: Part 1

Designation				Thickness			I	Z	r_y
Nominal size	*Mass per metre*	*Depth of section D*	*Width of section B*	*Web t*	*Flange T*	*Area of section*	*Moment of inertia Axis x–x*	*Elastic modulus Axis x–x*	*Radius of gyration Axis y–y*
(mm)	*(kg)*	*(mm)*	*(mm)*	*(mm)*	*(mm)*	*(cm^2)*	*(cm^4)*	*(cm^3)*	*(cm)*
203 × 152	52.09	203.2	152.4	8.9	16.5	66.4	4789	471.4	3.51
203 × 102	25.33	203.2	101.6	5.8	10.4	32.3	2294	225.8	2.25
178 × 102	21.54	117.8	101.6	5.3	9.0	27.4	1519	170.9	2.25
152 × 127	37.20	152.4	127.0	10.4	13.2	47.5	1818	238.7	2.82
152 × 89	17.09	152.4	88.9	4.9	8.3	21.8	881.1	115.6	1.99
152 × 76	17.86	152.4	76.2	5.8	9.6	22.8	873.7	114.7	1.63
127 × 114	29.76	127.0	114.3	10.2	11.5	37.3	979.0	154.2	2.55
127 × 114	26.79	127.0	114.3	7.4	11.4	34.1	944.8	148.8	2.63
127 × 76	16.37	127.0	76.2	5.6	9.6	21.0	569.4	89.66	1.70
127 × 76	13.36	127.0	76.2	4.5	7.6	17.0	475.9	74.94	1.72
114 × 114	26.79	114.3	114.3	9.5	10.7	34.4	735.4	128.6	2.54
102 × 102	23.07	101.6	101.6	9.5	10.3	29.4	486.1	95.72	2.29

A selection of joists from reference 7.1.

TABLE 7.3(b) Section properties: channels

CHANNELS
To BS4: Part 1

Designation		Depth of section D	Width of section B	Thickness		Area of Section	I Moment of inertia Axis $x-x$	Z Elastic modulus Axis $x-x$	r_y Radius of gyration Axis $y-y$
Serial size	Mass per metre			Web t	Flange T				
(mm)	(kg)	(mm)	(mm)	(mm)	(mm)	(cm^2)	(cm^4)	(cm^3)	(cm)
254 × 89	35.74	254.0	88.9	9.1	13.6	45.52	4448	350.2	2.58
254 × 76	28.29	254.0	76.2	8.1	10.9	36.03	3367	265.1	2.12
229 × 89	32.76	228.6	88.9	8.6	13.3	41.73	3387	296.4	2.61
229 × 76	26.06	228.6	76.2	7.6	11.2	33.20	2610	228.3	2.19
203 × 89	29.78	203.2	88.9	8.1	12.9	37.94	2491	245.2	2.64
203 × 76	23.82	203.2	76.2	7.1	11.2	30.34	1950	192.0	2.23
178 × 89	26.81	177.8	88.9	7.6	12.3	34.15	1753	197.2	2.66
178 × 76	20.84	177.8	76.2	6.6	10.3	26.54	1337	150.4	2.25
152 × 89	23.84	152.4	88.9	7.1	11.6	30.36	1166	153.0	2.66
152 × 76	17.88	152.4	76.2	6.4	9.0	22.77	851.5	111.8	2.24
127 × 64	14.90	127.0	63.5	6.4	9.2	18.98	482.5	75.99	1.88
102 × 51	10.42	101.6	50.8	6.1	7.6	13.28	207.7	40.89	1.48
76 × 38	6.70	76.2	38.1	5.1	6.8	8.53	74.1	19.46	1.12

A selection of channels from reference 7.1.

TABLE 7.4(a) Safe Load Tables.
Universal Beams

Nominal sizes	Span (m)	Safe total loads on steel beams for: Restrained compression flange (kN)	Unsestrained compression flange (kN)	Maximum deflection of 1/360 × span (kN)	Encased in concrete (kN)
305 × 165 × 54 kg	6.0	165	129	145	149
× 46 kg	6.0	142	82	123	123
× 40 kg	6.0	123	70	106	105
254 × 146 × 43 kg	5.0	133	99	117	132
× 37 kg	5.0	114	75	99	111
× 31 kg	5.0	93	55	79	82
254 × 102 × 28 kg	5.0	81	34	72	51
× 25 kg	5.0	70	26	61	39
× 22 kb	5.0	60	19	51	28
254 × 146 × 31 kg	4.0	116	82	116	116
254 × 102 × 25 kg	4.0	87	41	87	83
× 22 kg	4.0	74	30	74	71
203 × 133 × 30 kg	4.0	92	75	80	92
× 25 kg	4.0	76	55	66	76

Based on BS 499 and Grade 43 steel.

Beams simply supported with torsional end restraint.

Universal Columns

TABLE 7.4(b) Safe Load Tables.
Universal columns used as beams

Nominal sizes	Span (m)	*Safe total loads on steel beams for:* Restrained compression flange (kN)	Unrestrained compression flange (kN)	Maximum deflection of 1/360 × span (kN)	Encased in concrete (kN)
203 × 203 × 86 kg	6.0	187	187	117	187
× 71 kg	6.0	156	156	95	156
× 60 kg	6.0	128	128	75	128
× 52 kg	6.0	112	109	65	112
× 46 kg	6.0	99	94	56	99
203 × 203 × 86 kg	5.0	224	224	169	224
× 71 kg	5.0	187	187	137	187
× 60 kg	5.0	153	153	109	153
× 52 kg	5.0	134	134	94	134
× 46 kg	5.0	118	118	81	118

Used as beams

203 × 203 × 86 kg	4.0	281	281	265	281
× 71 kg	4.0	234	234	214	234
× 60 kg	4.0	191	191	170	191
× 52 kg	4.0	168	168	147	168
× 46 kg	4.0	148	148	127	148
152 × 152 × 37 kg	5.0	72	69	39	72
× 30 kg	5.0	58	55	31	58
× 23 kg	5.0	43	33	22	43
152 × 152 × 37 kg	4.0	90	90	62	90
× 30 kg	4.0	73	72	48	73
× 23 kg	4.0	54	50	35	54
152 × 152 × 37 kg	3.0	120	120	110	120
× 30 kg	3.0	97	97	86	97
× 23 kg	3.0	73	73	62	73

Based on BS 449 and Grade 43 steel.
Beams simply supported with torsional end restraint.

TABLE 7.4(c) Safe Load Tables
Channels and joists

Nominal sizes	*Span depth = span/20 (m)*	*Safe total loads on steel beams for:* Restrained compression flange (kN)	Unrestrained compression flange (kN)	Maximum deflection of 1/360 × span (kN)	Encased in concrete (kN)
203 × 152 × 52 kg	4.0	155	155	134	155
102 × 25 kg	4.0	74	44	64	66
178 × 102 × 21 kg	3.5	64	45	55	64
152 × 127 × 37 kg	3.0	155	105	90	155
89 × 17 kg	3.0	51	39	44	51
76 × 18 kg	3.0	50	36	43	50
127 × 114 × 30 kg	2.5	81	81	70	81
× 27 kg	2.5	78	78	67	78
76 × 16 kg	2.5	47	43	41	47
× 13 kg	2.5	39	33	34	39
114 × 114 × 27 kg	2.5	68	68	53	68
102 × 102 × 23 kg	2.0	63	63	54	63
× 64 × 9 kg	2.0	28	25	24	28
× 44 × 7 kg	2.0	20	11	17	16

Channels

254 × 89 × 36 kg	5.0	92	54	80	81
× 76 × 28 kg	5.0	70	31	60	46
229 × 89 × 33 kg	4.5	87	66	75	82
× 76 × 26 kg	4.5	67	36	58	54
203 × 89 × 30 kg	4.0	81	68	70	80
× 76 × 24 kg	4.0	63	44	55	60
178 × 89 × 27 kg	3.5	74	70	64	74
× 76 × 21 kg	3.5	57	44	49	57
152 × 89 × 24 kg	3.0	67	67	58	67
× 76 × 18 kg	3.0	49	43	42	49
127 × 64 × 15 kg	2.5	40	38	34	40
102 × 51 × 10 kg	2.0	27	25	23	27
76 × 38 × 7 kg	1.5	17	17	15	17

Based on BS 449 and Grade 43 steel.
Beams simply supported with torsional end restraint.

TABLE 7.5 Allowable stress in bending: mild steel (grade 43)

	D/T							
l/r_y	*10*	*15*	*20*	*25*	*30*	*35*	*40*	*50*
90	165	165	165	165	165	165	165	165
95	165	165	165	163	163	163	163	163
100	165	165	165	157	157	157	157	
105	165	165	160	152	152	152	152	152
110	165	165	156	147	147	147	147	147
115	165	165	152	141	141	141	141	141
120	165	162	148	136	136	136	136	136
130	165	155	139	126	126	126	126	126
140	165	149	130	115	115	115	115	115
150	165	143	122	104	104	104	104	104
160	163	136	113	95	94	94	94	94
170	159	130	104	91	85	82	82	82
180	155	124	96	87	80	76	72	71
190	151	118	93	83	77	72	68	62
200	147	111	89	80	73	69	64	59
210	143	105	87	77	70	65	61	55
220	139	99	84	74	67	62	58	52
230	134	95	81	71	64	59	55	49
240	130	92	78	69	61	56	52	47
250	126	90	76	66	59	54	50	44

After Table 3a, BS 449.

The tables in this chapter have been based on BS 449 which is an elastic method of design. Structural Engineers would normally use BS 5950, which is a limit state method as explained on page 9.

REFERENCES

7.1 *Structural Steel Sections*, British Standard 4: Part 1.
7.2 *Structural Use of Steel*, British Standard 5950: Part 1: 1985.
7.3 *Specification for the Use of Structural Steel in Building*, British Standard 449: Part 2.
7.4 *Handbook on Structural Steel*, BCSA/SCI.

Worked Example 7.1
Project: Steel beams
Restrained and unrestrained compression flanges

Ref.	Calculations	Output
	Question. A steel beam is to span 4m. Assume the beam is simply supported and has adequate torsional restraint at its ends. a) What size of beam is required to support a total uniform load of 78 kN, assuming that the top flange has adequate lateral bracing? b) What load will the beam selected in (a) support, if the top flange has no lateral bracing?	
Table 1·1	Approximate depth for 4 m span $\text{depth} = \frac{\text{span}}{20} = 200 \text{ mm}$ $\text{depth} = \frac{\text{span}}{18} = 225 \text{ mm}$	
	a) First examine the safe load tables for UB, Joists and channels. The safe load has to be greater than the actual total load of 78 kN.	
Table 7·4	For Universal Beam, span 4m UB 254 × 102 × 25 kg, safe load = 87 kN (Also deflection load = 87 kN)	span 4 m
	UB 203 × 133 × 30 kg, safe load = 92 kN (And deflection load = 80 kN)	
	Joist 203 × 102 × 25 kg, safe load = 74 kN	Joist no good
	channel 203 × 89 × 30 kg, Safe load = 81 kN (but deflection load is only = 70 kN)	Deflection no good

Worked Example 7.1 continued

Ref.	Calculations	Output
	Deflection. Check deflection loads given in table 7·5	
Table 7·2	UB 203 × 133 × 30 kg I = 2880 cm^4 C = 28·0 (4m span) Deflection Load = $W_D = \frac{CI}{1000}$ = 80 kN	Deflection OK.
Table 7·3	Channel 203 × 89 × 30 kg I = 2491 cm^4 C = 28·0 (4m span) Deflection Load = $W_D = \frac{CI}{1000}$ = 70 kN (No good)	Excessive deflection
	Beam selection The lightest steel beam which meets the requirements of strength and deflection is therefore: UB 254 × 102 × 25kg	
	b) Beam without lateral support. Consider UB 254 × 102 × 25 kg. As the beam has no lateral bracing either use table 7·5 for unrestrained compression flange or work out from first principles. From first principles Laws of bending $f = \frac{M}{Z}$ where f = stress M = bending moment Z = elastic modulus	

Worked Example 7.1 continued

Ref.	Calculations	Output
Worked Example 3.3	W = max. load the beam can carry $M = \frac{WL}{8} = \frac{W \times 4m}{8} = 0.5\,W$ kN.m	
	Properties of beam $z = 265\,000$ mm^3 $r_y = 21.1$ mm $D = 257$ mm $T = 8.4$ mm $D/T = 30.6$ $L = 4000$ mm $L/r_y = 190$	
BS 449 Table 3a	Therefore permissible stress = 77 N/mm^2 Actual stress = $f = M/Z$ If Actual stress = Permissible stress $M/Z = 77$ N/mm^2 $\frac{500\,000\,W \text{ N.mm}}{265000 \text{ mm}^3} = 77$ N/mm^2 Total uniform load W = 41 kN	Max. Uniform Load=41kN
	In this example, the safe maximum uniform load is less, if the top compression flange is not restrained For UB 254 × 102 × 25 kg Restrained flange Safe load = 87 kN Unrestrained flange Safe load = 41 kN Deflection limited to 1/360 span Safe load = 87 kN	

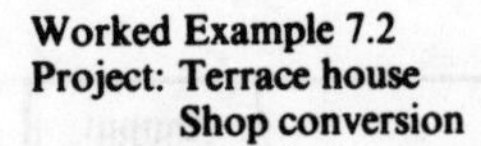
Worked Example 7.2
Project: Terrace house
Shop conversion

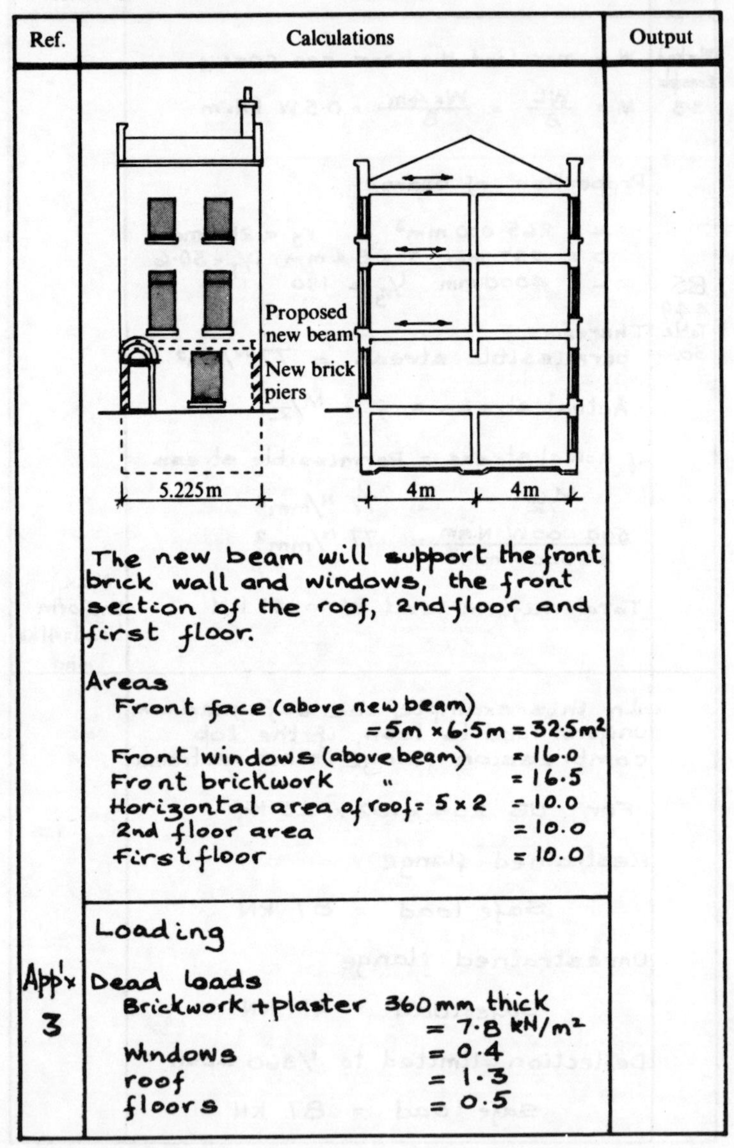

Ref.	Calculations	Output
	The new beam will support the front brick wall and windows, the front section of the roof, 2nd floor and first floor.	
	Areas	
	Front face (above new beam) $= 5\text{m} \times 6{\cdot}5\text{m} = 32{\cdot}5\text{m}^2$	
	Front windows (above beam) $= 16{\cdot}0$	
	Front brickwork " $= 16{\cdot}5$	
	Horizontal area of roof $= 5 \times 2 = 10{\cdot}0$	
	2nd floor area $= 10{\cdot}0$	
	First floor $= 10{\cdot}0$	
	Loading	
App'x 3	Dead loads	
	Brickwork + plaster 360 mm thick $= 7{\cdot}8\ \text{kN/m}^2$	
	Windows $= 0{\cdot}4$	
	roof $= 1{\cdot}3$	
	floors $= 0{\cdot}5$	

Worked Example 7.2 continued

Ref.	Calculations	Output
App'x 2	Imposed loads roof (23° pitch) = 0·75 kN/m² 2nd floor = 1·5 First floor = 2·5	
	Load on beam (dead) Brickwork + plaster = 16·5 m² × 7·8 kN/m² = 129·0 kN Windows = 16 m² × 0·4 = 6·4 roof = 10 m² × 1·3 = 13·0 2nd floor = 10 m² × 0·5 = 5·0 First floor = 10 m² × 0·5 = 5·0 158·4 kN	
App'x 2 table 5	Load on beam (Imposed) roof = 10m² × 0.75 kN/m² = 7.5 kN 2nd floor = 10m² × = 15.0 First floor = 10m² × = 25.0 47·5 10% reduction for 2 floors = 4·7 42·8 kN Total load = 158·4 + 42·8 = 201 kN	Total load 201 kN
table 7·4	Safe load tables span 5m. Use 2no UB 305 × 127 × 37Kg Safe load = 2 × 124 = 248 kN (deflection OK) Lateral restraint Bolt or weld the beams together to get sufficient bracing for flange.	New Beam 2no UB 305 × 127 × 37 Kg.

Worked Example 7.3
Project: Cottage extension
Steel beam over window

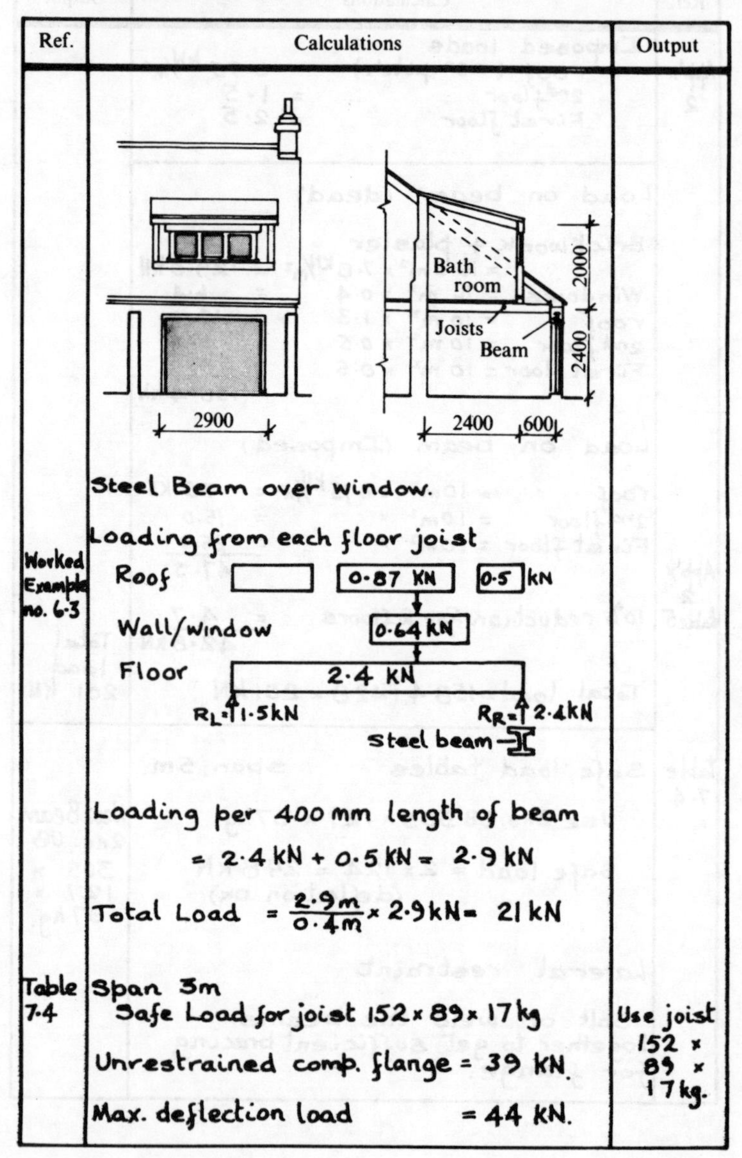

Ref.	Calculations	Output
	Steel Beam over window.	
	Loading from each floor joist	
Worked Example no. 6·3	Roof: 0·87 KN, 0·5 KN	
	Wall/window: 0·64 KN	
	Floor: 2·4 KN	
	R_L = 1·5 kN; R_R = 2·4 kN; steel beam	
	Loading per 400 mm length of beam = 2·4 kN + 0·5 kN = 2·9 kN	
	Total Load $= \frac{2.9\,m}{0.4\,m} \times 2.9\,kN = 21\,kN$	
Table 7·4	Span 3m	
	Safe Load for joist 152 × 89 × 17 kg	Use joist 152 × 89 × 17 kg.
	Unrestrained comp. flange = 39 KN.	
	Max. deflection load = 44 KN.	

Worked Example 7.4
Project: Shop extension
Steel beams

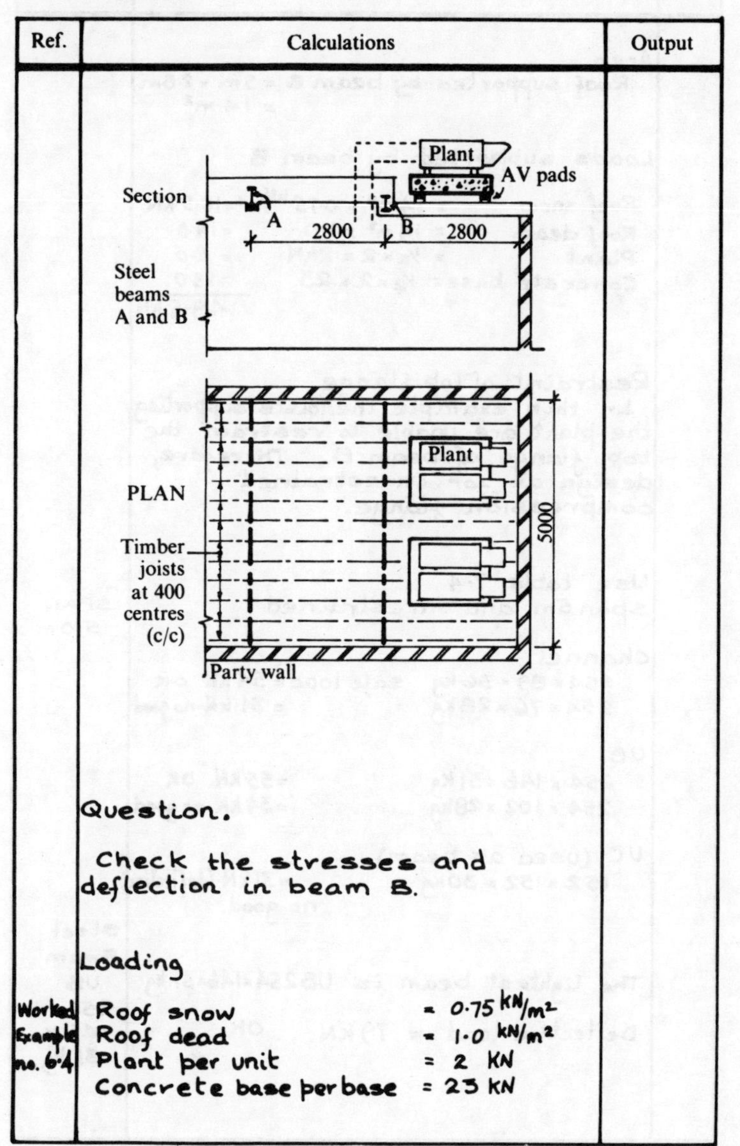

Question.

Check the stresses and deflection in beam B.

Loading

Ref.	Calculations		Output
Worked Example no. 6·4	Roof snow	= 0·75 kN/m²	
	Roof dead	= 1·0 kN/m²	
	Plant per unit	= 2 kN	
	Concrete base per base	= 23 kN	

Worked Example 7.4 continued

Ref.	Calculations	Output
	Area Roof supported by beam B = 5m × 2·8m = 14 m²	
	Loads supported by beam B Roof snow = 14m² × 0·75 kN/m² = 10·5 kN Roof dead = 14m² × 1·0 = 14·0 Plant = ½ × 2 × 2kN = 2·0 Concrete base: ½ × 2 × 23 = 23·0 49·5 kN	
	Restraint of top flange In this example, the joists supporting the plant are unable to restrain the top flange of beam B. Therefore, design as for unrestrained compression flange.	
	Use table 7·4 span 5m and unrestrained	SPAN 5·0m
	Channel 254 × 89 × 36kg safe load = 54 kN OK 254 × 76 × 28kg = 31 kN no good	
	UB 254 × 146 × 31kg = 55 kN OK 254 × 102 × 28kg = 34 kN no good	
	UC (used as beam) 152 × 152 × 30kg = 31kN (deflection) no good.	
	The lightest beam is UB 254 × 146 × 31 kg Deflection load = 79 kN OK.	Steel Beam UB 254 × 146 × 31 kg.

CHAPTER 8

BRICK AND BLOCK WALL DESIGN

There are a large variety of bricks and a smaller variety of blocks on the market. There are also a variety of mortars, which can be used to bond these bricks and blocks together.

In general the bricks and blocks can be divided into:

(*a*) Clay bricks
(*b*) Calcium silicate bricks
(*c*) Concrete bricks
(*d*) Clay blocks
(*e*) Dense concrete blocks
(*f*) Aerated (lightweight) concrete blocks

The variety of mortars are:

(*a*) Cement sand mortar
(*b*) Cement lime sand
(*c*) Masonry cement and sand
(*d*) Cement sand with plasticiser
(*e*) Lime and sand

To ascertain the strength of a wall, it is necessary to know what bricks, blocks and mortars are to be used. The choice of bricks and blocks is usually a visual decision rather than a strength requirement, and should be made at an early stage to enable the designer to carry out the masonry calculations.

8.1 BRICK AND BLOCK STRENGTH

Figure 8.1 shows the relationship between the crushing strength of a variety of bricks, the mortar strength and the characteristic strength of masonry used to calculate the strength of walls. Larger units, such as blocks, are not included in this diagram for reasons connected with shape factor, as discussed later in this chapter. However, the diagram does include both concrete and clay products. The actual crushing strength is obtained from the manufacturer, and the strength of the mortar will depend upon the mix proportions of cement, lime and sand. From these strengths, the characteristic strength is obtained from either Tables 8.1, 8.2 and 8.3 or from references 8.1 and 8.2.

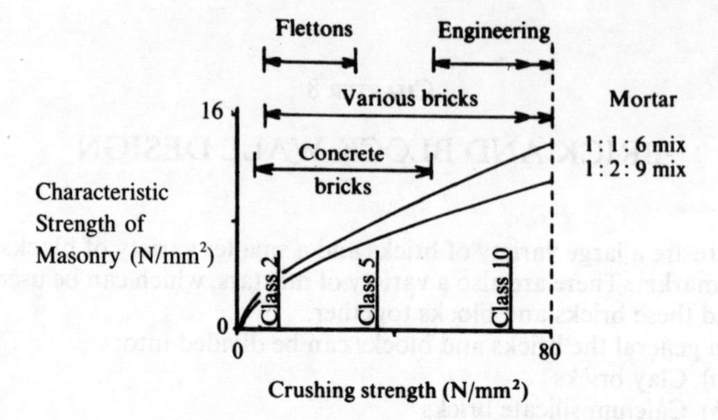

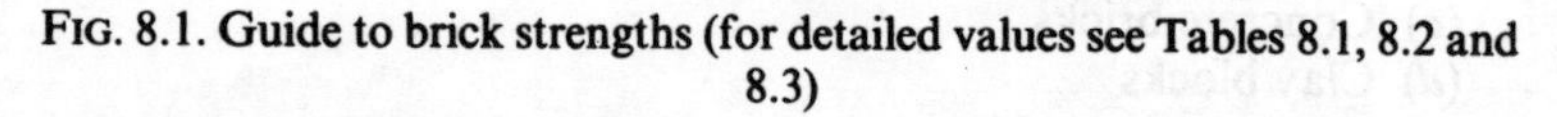
FIG. 8.1. Guide to brick strengths (for detailed values see Tables 8.1, 8.2 and 8.3)

Crushing strength

The manufacturers measure the compressive crushing strength on the gross area of the brick or block, regardless of whether the unit is solid or hollow. So the quoted stresses are not necessary related to the quality of the material used, but to the area of voids in the unit. Therefore:

$$\text{Crushing stress} = \frac{\text{Ultimate crushing force on one unit}}{\text{Length} \times \text{width of unit}}$$

Mortar grade

The choice of mortar will depend upon a combination of strength requirement, and flexibility of the bedding joint to accommodate movement. As would be expected, stronger mortars produce stronger masonry, but this extra strength is at the expense of flexibility, which decreases as the strength of the mortar increases. Brick and blockwork must be considered as a material constantly on the move, reacting to the changing atmospheric conditions of moisture content and temperature as well as to any movement in the foundations. To achieve flexibility, it is therefore often appropriate to specify a weaker mortar rather than a very strong one.

There are four structural grades of mortars, (i), (ii), (iii) and (iv). Grade (i) is most appropriate to the compressive strength of engineering bricks, and grade (ii) to high-quality clay bricks and top-quality dense concrete blocks. For strengths below this, it is appropriate to use grades (iii) and (iv) (see Table 8.2 on p. 134 at the end of this chapter). This chapter only deals with the lower grades of

bricks and blocks, and therefore, only with the lower grades of mortars.

Shape factor of unit

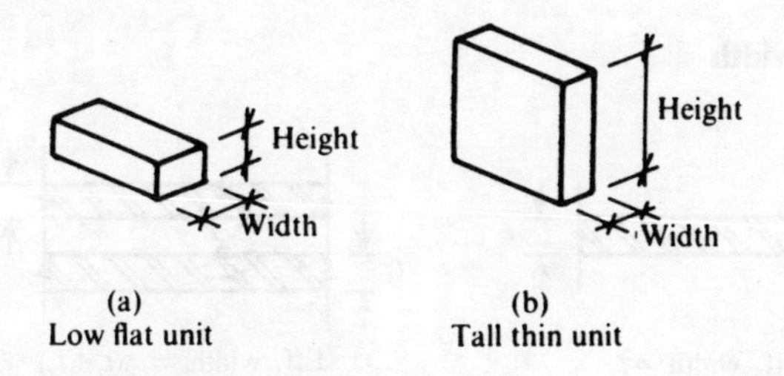

FIG. 8.2. Dimensional shape of units

The actual strength of a brick or block unit will be influenced by the dimensional shape of the unit (Fig. 8.2). For the lower compressive strengths, the testing machine gives a misleading value as the units become thinner and taller. For example, a unit with a crushing strength of 2.8N/mm^2 and a height to width ratio of 2, will have twice the strength of a unit of the same crushing strength but with a ratio of height to width of only 0.6. The enhanced strength is lost if the blocks are hollow.

8.2 SLENDERNESS RATIOS AND MODIFICATION FACTORS

The slenderness ratio of a wall and the eccentricity of the load will indicate whether the wall will crush or buckle. The greater the slenderness and eccentricity, the sooner buckling will occur, so the strength of the wall will depend upon these two factors.

In order to calculate the slenderness ratio, it is necessary to know the effective length or effective height of the wall and to know its effective thickness.

$$\text{Slenderness ratio} = \frac{\text{Effective length or height}}{\text{Effective width}}$$

Effective length

The effective length or height of a wall depends upon how the wall is held at its ends or at points of lateral support. The wall can either be considered to span in a vertical direction, such as between two floors, or it can span in a horizontal direction as between two returns. This is discussed in more detail in section 8.3, and at this stage it is only necessary to appreciate that a wall can span either in the horizontal direction or the vertical direction.

Effective width

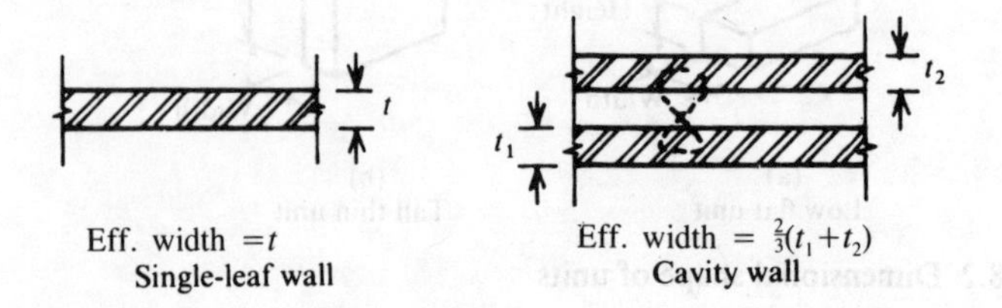

FIG. 8.3. Effective thickness of walls

The effective width of a single-leaf wall will be the width of the wall (Fig. 8.3). If it is a cavity wall, then a 'guess' has to be made as to its effective width. The 'guess' used in reference 8.1 is $\frac{2}{3}(t_1+t_2)$, where t_1 and t_2 are the thickness of the two leaves. Other guesses are made when walls are stiffened by piers, or curved in plan. It must be said here that the 'guesses' are backed up with experimental and practical experience.

Once the effective length, height and width have been found, and it has been established whether the wall has horizontal or vertical lateral support, then the slenderness ratio of the wall can be calculated.

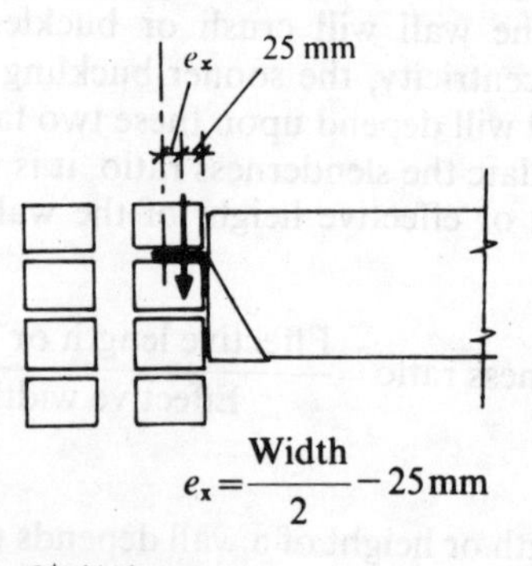

$$e_x = \frac{\text{Width}}{2} - 25\,\text{mm}$$

FIG. 8.4. Eccentricity of joist hanger

The next step is to ascertain the eccentricity of the load at the top of the wall. For example, joist hangers are considered to be supported by the wall, 25 mm in from the face of the wall (Fig. 8.4), and will produce an eccentric load on the wall, so that:

Eccentricity of load $= e_x = 0.5t - 25\text{mm}$

where t = thickness of the wall.

If the eccentricity of the load acts within 5 per cent of the centre of the wall, it is considered to be axially loaded, or in other words loaded at its centre.

The reason why it is necessary to know the eccentricity of the load, is that the eccentricity will increase the tendency for the wall to buckle in the same way as the slenderness does. Table 8.3 (p. 135) caters for which are called capacity reduction factors in reference 8.1. These factors reduce the characteristic strength of the masonry.

8.3 PRACTICAL CONSIDERATIONS OF STABILITY

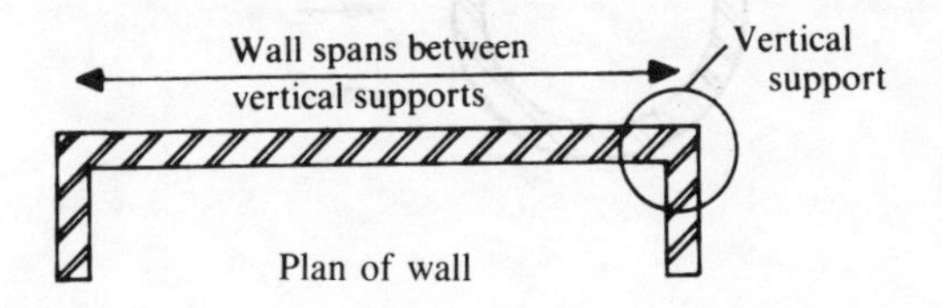

FIG. 8.5

Overall stability of a masonry structure can easily be achieved by a sensible plan arrangement of the walls, as illustrated in Fig. 8.5. This is especially so for higher buildings where the lateral effect of the wind becomes an important factor. The horizontal force of the wind has to be transmitted through the walls down to the foundations without producing any tension in the walls. Any tension the wind might produce in the walls must be cancelled by the compression produced by the dead weight of the walls and the structure as a whole.

So how is the wind load transmitted? This is best explained by considering the wind blowing on to the front face of a building. The front wall itself will have little strength to resist the wind. What has to happen is for the load to be transferred to the cross walls which are, say, perpendicular to the front wall. This can be achieved in two ways, either by the cross walls coming into contact with the front wall and giving it vertical support right up the building or by each floor acting as a diaphragm transferring the force from the front wall to the cross walls, and providing horizontal stiff supports to the front wall. In practice, of course, both these methods can work together and to

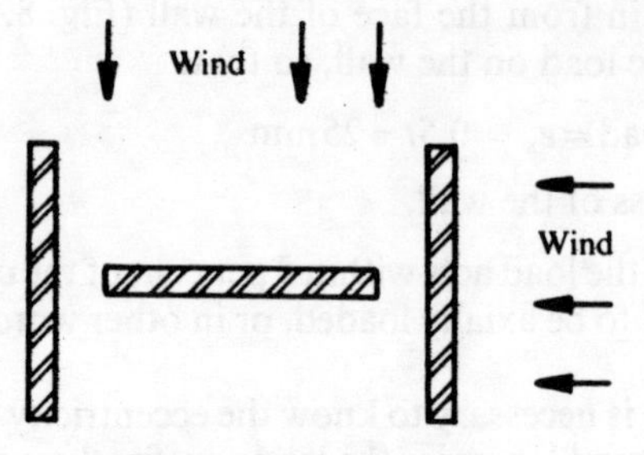

Shear walls at right angles

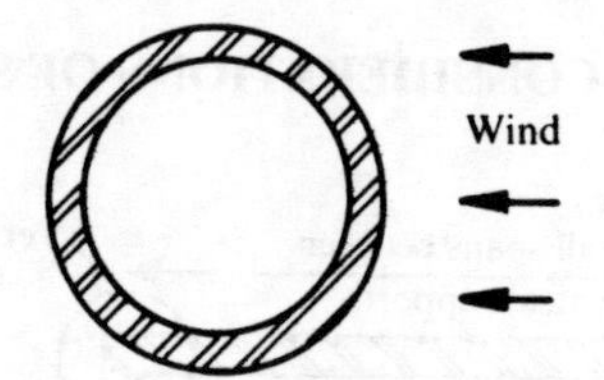

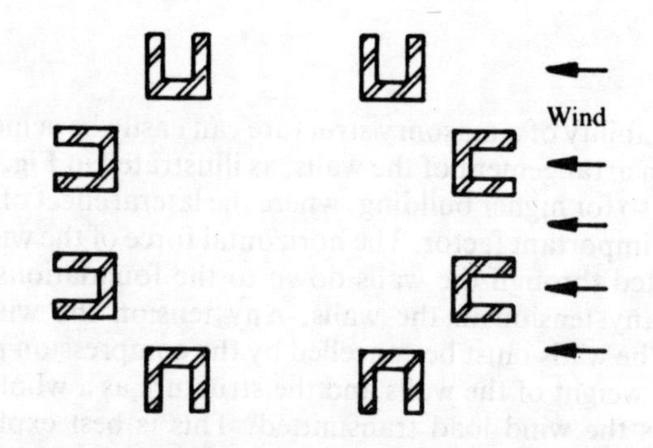

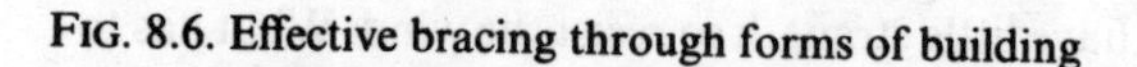

FIG. 8.6. Effective bracing through forms of building

what extent each contributes will depend upon the layout of the plan of the supporting walls (Fig. 8.5). Once the load has been transferred to the cross walls, it can easily be transferred to the foundations.

The wall panels themselves have also to be supported and have sufficient stiffness not to buckle. These panels or walls can either have vertical support as in Fig. 8.6 or horizontal support as in Fig. 8.7. The vertical supports are either generated by cross walls as discussed above or by a change in the direction of the wall as in Fig. 8.6, which

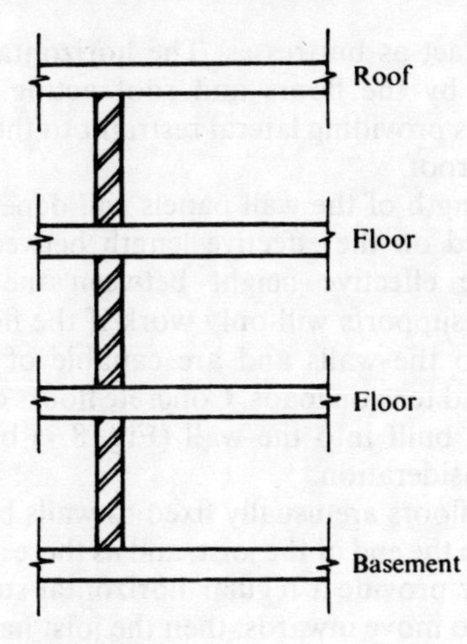

FIG. 8.7. Wall restrained by floors

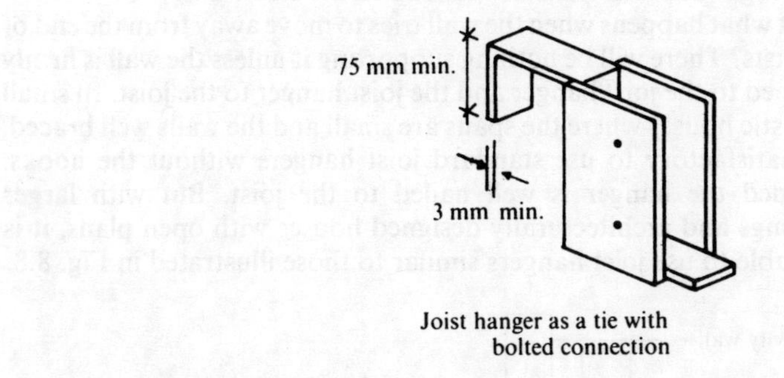

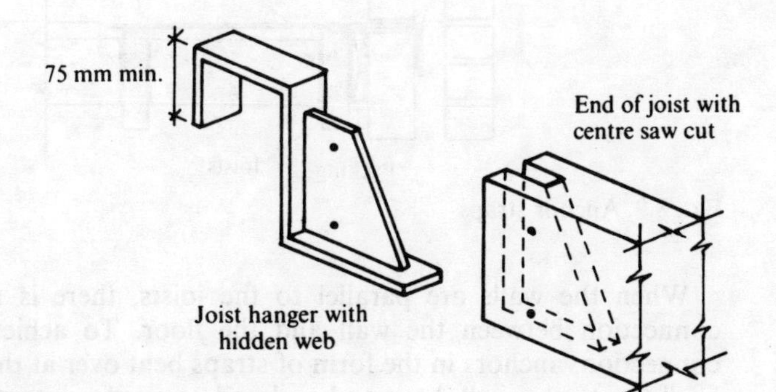

FIG. 8.8. Joist hangers

in a sense act as buttresses. The horizontal supports are generally developed by the floors and roof acting as very stiff horizontal diaphragms providing lateral restraint to the walls at each floor level and at the roof.

The strength of the wall panels will depend upon the slenderness ratios based on the effective length between the vertical supports and/or the effective height between the floors and roof. The horizontal supports will only work if the floors and roof are firmly attached to the walls and are capable of transferring both compression and tension loads. Concrete floors do not cause any concern if they are built into the wall (Fig. 8.7) but timber floors require careful consideration.

Timber floors are usually fixed to walls by means of joist hangers attached to the end of the joist, and as these are usually about 400 mm apart, they provide a regular horizontal support to the wall. If the wall tries to move inwards, then the joist hangers are quite adequate to support the weight of the wall. The floor, being constructed of floorboards, joists and ceiling plaster, will act as the horizontal diaphragm and will transfer the load to the side walls.

But what happens when the wall tries to move away from the end of the joists? There will be nothing supporting it unless the wall is firmly attached to the joist hanger and the joist hanger to the joist. In small domestic houses where the spans are small and the walls well braced, it is satisfactory to use standard joist hangers without the hooks, provided the hanger is well nailed to the joist. But with larger buildings and architecturally designed houses with open plans, it is advisable to use joist hangers similar to those illustrated in Fig. 8.8.

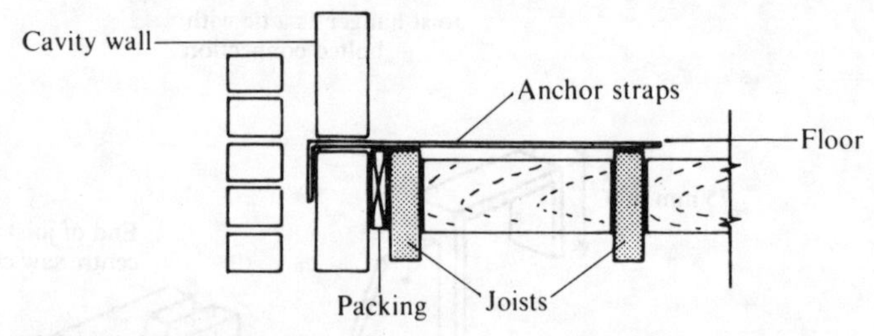

FIG. 8.9. Anchor straps

When the walls are parallel to the joists, there is no natural connection between the wall and the floor. To achieve a good connection, anchors in the form of straps bent over at their ends to hook on to the wall, have to be placed at regular intervals and be strong enough to transmit the lateral loads (Fig. 8.9). It is recommended that the anchors are placed not more than 2 m apart in houses and 1.25 m apart in other buildings. The cross-sectional dimensions of the anchors should be at least 30 mm × 5 mm.

8.4 DESIGN FOR MOVEMENT

The dimensions of bricks, blocks and their mortars will change depending on their moisture content and temperature. It is important to produce details which allow for this movement without causing damage to the brick and blockwork or to neighbouring materials. It is possible to estimate the amount of movement to be expected by using the information given in Appendix C of reference 8.3.

To avoid any damage, it is advisable to introduce movement joints in positions which fit into the design of the building. They should be considered at the earliest stages and be part of the initial design. As a general rule, it is recommended to have a 10 mm wide vertical movement joint every 12 m for clay bricks, 8 m for calcium silicate bricks and 6 m for concrete blocks. These joints are both expansion and contraction joints.

Cracking is most likely to occur at sections where the vertical or horizontal section of the wall changes abruptly, such as openings for windows and doorways. Concrete blocks are especially sensitive at these weakened sections. The concrete blocks like to have plenty of height to produce high compression stresses in the wall which discourages cracking. The weakened sections are areas where the compressive stress is low, such as panels under windows, even if their lengths are less than 6 m. Cracking can be controlled in these weakened areas by using reinforcement in the horizontal mortar bed. Other points of weakness, such as over and around windows, can also be controlled by the use of bed reinforcement.

8.5 STRUCTURAL DESIGN OF MASONRY LIMIT STATE DESIGN METHOD

In the chapters covering timber and steel, the elastic method of structural design has been used. In this chapter and the one on concrete, the limit state method is used. The two methods were discussed in section 1.3 and it is advisable to refer to that section and also to sections 9.1 to 9.3 on the structural design of concrete.

8.6 GENERAL REQUIREMENTS OF LIMIT STATE DESIGN

Limit state design considers all aspects of ultimate safety and functional requirements. Each aspect is called a limit state, so when designing a masonry structure, all the possible limit states should be considered. The structure must be safe, it must be serviceable in that it must function properly with no cracking which will impair the visual finish. It must, of course, be stable, durable and meet the

required fire resistance. Each limit state has to be dealt with separately, and each one will have to meet the minimum requirements. If one limit state is not met, then the whole design will fail.
The various limit states are:

1. Strength
2. Stability
3. Serviceability (so a building can function during its lifetime)

8.7 LOADS AND SAFETY FACTORS

Design Load = Safety factor × Characteristic load

The characteristic load is the actual load or the estimated likely load to be resisted by the structure. The partial safety factor will depend upon the combination of loads and how they may affect the structure. For dead and imposed only:

Design dead load = 1.4 × Characteristic dead load.
Design imposed load = 1.6 × Characteristic imposed load.

8.8 CHARACTERISTIC COMPRESSIVE STRENGTH

The characteristic compressive strength for normally bonded masonry may be taken from the values given in table 8.1. These values are defined in terms of the shape, compressive strength of the structural units and the designation of the mortar. Linear interpolation within the tables is permitted.

8.9 DESIGN STRENGTH OF MASONRY

If normal manufacturing and construction control is assumed, then the partial safety factor for the masonry is equal to 3.5 as given in reference 8.1. Then:

$$\text{Design Strength} = \frac{\text{Capacity reduction factor}}{3.5} \times \text{Characteristic strength}$$

TABLE 8.1. Characteristic compressive strength of masonry

(a) Constructed with standard format bricks

Mortar designation	Compressive strength of unit (N/mm²)							
	5	10	15	20	27.5	35	50	70
(iii)	2.5	4.1	5.0	5.8	7.1	8.5	10.6	13.1
(iv)	2.2	3.5	4.4	5.2	6.2	7.3	9.0	10.8

(b) Constructed with blocks having a ratio of height to least horizontal dimension of 0.6

Mortar designation	Compressive strength of unit (N/mm²)							
	2.8	3.5	5.0	7.0	10	15	20	35 or greater
(iii)	1.4	1.7	2.5	3.2	4.1	5.0	5.8	8.5
(iv)	1.4	1.7	2.2	2.8	3.5	4.4	5.2	7.3

(c) Constructed with hollow blocks having a ratio of height to least horizontal dimension of between 2.0 and 4.0

Mortar designation	Compressive strength of unit (N/mm²)							
	2.8	3.5	5.0	7.0	10	15	20	35 or greater
(iii)	2.8	3.5	5.0	5.4	5.5	5.7	5.9	8.5
(iv)	2.8	3.5	4.4	4.8	4.9	5.1	5.3	7.3

(d) Constructed from solid concrete blocks having a ratio of height to least horizontal dimension of between 2.0 and 4.0

Mortar designation	Compressive strength of unit (N/mm²)							
	2.8	3.5	5.0	7.0	10	15	20	35 or greater
(iii)	2.8	3.5	5.0	6.4	8.2	10.0	11.6	17.0
(iv)	2.8	3.5	4.4	5.6	7.0	8.8	10.4	14.6

After Table 2, BS 5628: Part 1.

TABLE 8.2 Mortar strength

Mortar designation	*Mortar mix*			*Mean compression strength at 28 days*	
	Cement: lime:sand	*Masonry cement:sand*	*Cement:sand with plasticiser*	*Preliminary tests*	*Site tests*
(iii)	1:1:6	1:4	1:6	3.6	2.5
(iv)	1:2:9	1:6	1:8	1.5	1.0

After Table 1, BS 5628: Part 1.

TABLE 8.3 Capacity reduction factors

Slenderness	Eccentricity at top of wall t = thickness of wall			
Effective height / *Effective width*	*Axially loaded or up to 0.05*t	*0.1*t	*0.2*t	*0.3*t
0	1.00	0.88	0.66	0.44
6	1.00	0.88	0.66	0.44
8	1.00	0.88	0.66	0.44
10	0.97	0.88	0.66	0.44
12	0.93	0.87	0.66	0.44
14	0.89	0.83	0.66	0.44
16	0.83	0.77	0.64	0.44
18	0.77	0.70	0.57	0.44
20	0.70	0.64	0.51	0.37
22	0.62	0.56	0.43	0.30
24	0.53	0.47	0.34	
26	0.45	0.38		
27	0.40	0.33		

After Table 7, BS 5628: Part 1.

REFERENCES

8.1 *Code of Practice for Use of Masonry,* British Standard 5628: Part 1: 1978, structural use of unreinforced masonry.
8.2 *Code of Practice for Use of Masonry,* British Standard 5628: Part 2: 1985, structural use of reinforced and prestressed masonry.
8.3 *Code of Practice for Use of Masonry,* British Standard 5628: Part 3: 1985, materials and components, design and workmanship.

Worked Example 8.1
Project: Studio house
Calculation of stresses in rear wall

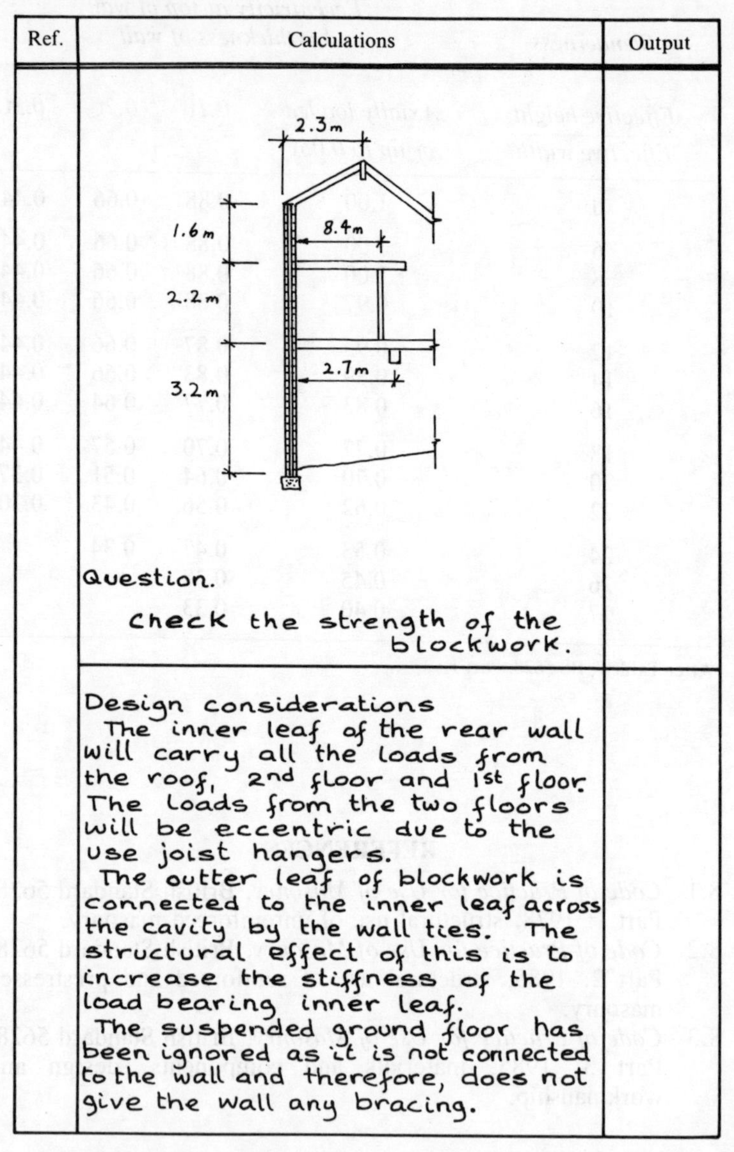

Ref.	Calculations	Output
	Question. Check the strength of the blockwork.	
	Design considerations The inner leaf of the rear wall will carry all the loads from the roof, 2nd floor and 1st floor. The loads from the two floors will be eccentric due to the use joist hangers. The outter leaf of blockwork is connected to the inner leaf across the cavity by the wall ties. The structural effect of this is to increase the stiffness of the load bearing inner leaf. The suspended ground floor has been ignored as it is not connected to the wall and therefore, does not give the wall any bracing.	

Worked Example 8.1 continued

Ref.	Calculations	Output
Manufacturer	**Dimensions and strength.** Co-ordinating face dimensions 450 × 225 Block size 440 × 215 × 100 mm Cavity 75 mm strength $2.8\ N/mm^2$	
App'x 2 and 3	**Loading.** Imposed roof $0.75\ kN/m^2$ floor 1.5 Dead roof 0.8 floor 0.5 Wall (each leaf) 1.4	
	Areas supported by one metre run of wall. Roof $= \frac{1m \times 2.3m}{2} = 1.15 m^2$ 2nd floor $= \frac{1m \times 2.4m}{2} = 1.20 m^2$ 1st floor $= \frac{1m \times 2.7m}{2} = 1.35 m^2$	
	Loads on inner leaf Roof: LL = 1.15 × 0.75 = 0.86 kN; DL = 1.15 × 0.80 = 0.92 kN Wall 1.6 × 1.4 = 2.24 kN 2nd fl.: LL = 1.2 × 1.5 = 1.80 kN; DL = 1.2 × 0.5 = 0.60 kN Wall 2.2 × 1.4 = 3.08 kN 1st fl.: LL = 1.35 × 1.5 = 2.03 kN; DL = 1.2 × 0.5 = 0.68 kN Wall 3.2 × 1.4 = 4.48 kN G.L.	

Worked Example 8.1 continued

Ref.	Calculations	Output
	Load on inner leaf of wall. a) second floor level Dead = 0·92 + 2·24 + 0·6 = 3·76 kN Imposed = 0·86 + 1·80 = 2·66 kN 6·42 kN b) First floor level Dead = 3·76 + 3·08 + 0·68 = 7·52 kN Imposed = (2·66 + 2·03) less 10% = 4·22 kN 11·74 kN c) Ground Level Dead = 7·52 + 4·48 = 12·00 kN Imposed = 4·22 = 4·22 kN 16·22 kN per m length of wall.	
	Design dead load = 1·4 × Characteristic load Design imposed Load = 1·6 × characteristic load Total Design Load: Level (a) 1·4 × 3·76 kN + 1·6 × 2·66 kN = 9·52 kN. (b) 1·4 × 7·52 kN + 1·6 × 4·22 kN = 17·28 kN. (c) 1·4 × 12·00 kN + 1·6 × 4·22 kN = 23·55 kN.	
	Eccentricity of Load ℄ Axial Load First floor Load x Resultant Load 25 mm First floor level Axial load = 6·42 + 3·08 = 9·5 kN Ecc. load = 2·03 + 0·68 = 2·7 kN 12·2 kN.	

Worked Example 8.1 continued

Ref.	Calculations	Output
	$x = \frac{25\,mm \times 2.7\,kN}{12.2\,kN} = 5\,mm$	
	Eccentricity of resultant $= \frac{5\,mm}{100\,mm} = 0.05$	
	Therefore, the wall is considered to be axially loaded.	
Fig. 8.3	Effective width $= 2/3(t_1 + t_2)$ $= 2/3(100 + 100) = 133$ mm	
	Effective height	
Table 8.3	Wall / height / effective height / SR / Capacity Reduction factor 2nd / 1600 / 2h = 3200 / 24 / 0.53 1st / 2200 / h = 2200 / 16.5 / 0.81 Ground / 3200 / h = 3200 / 24 / 0.53	
Table 8.1 (d)	If the inner leaf is constructed with 2.8 N/mm², 440 × 215 × 100 mm solid concrete blockwork (h/width = 2.15) with 1:2:8 mortar, then: Compressive strength = 2.8 N/mm²	
	Design strength $= \frac{\text{Capacity R.F.}}{3.5} \times 2.8\,N/mm^2 \times 1000 \times 100\,mm$ = 80 kN × Capacity R.F.	
	At Level (a) Design strength = 80 kN × 0.53 = 42 kN. (b) " = 80 kN × 0.81 = 64 kN. (c) " = 80 kN × 0.53 = 42 kN.	Strength OK.

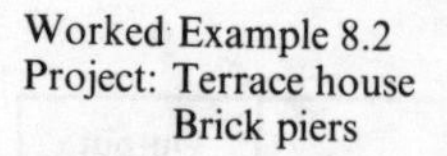
Worked Example 8.2
Project: Terrace house
Brick piers

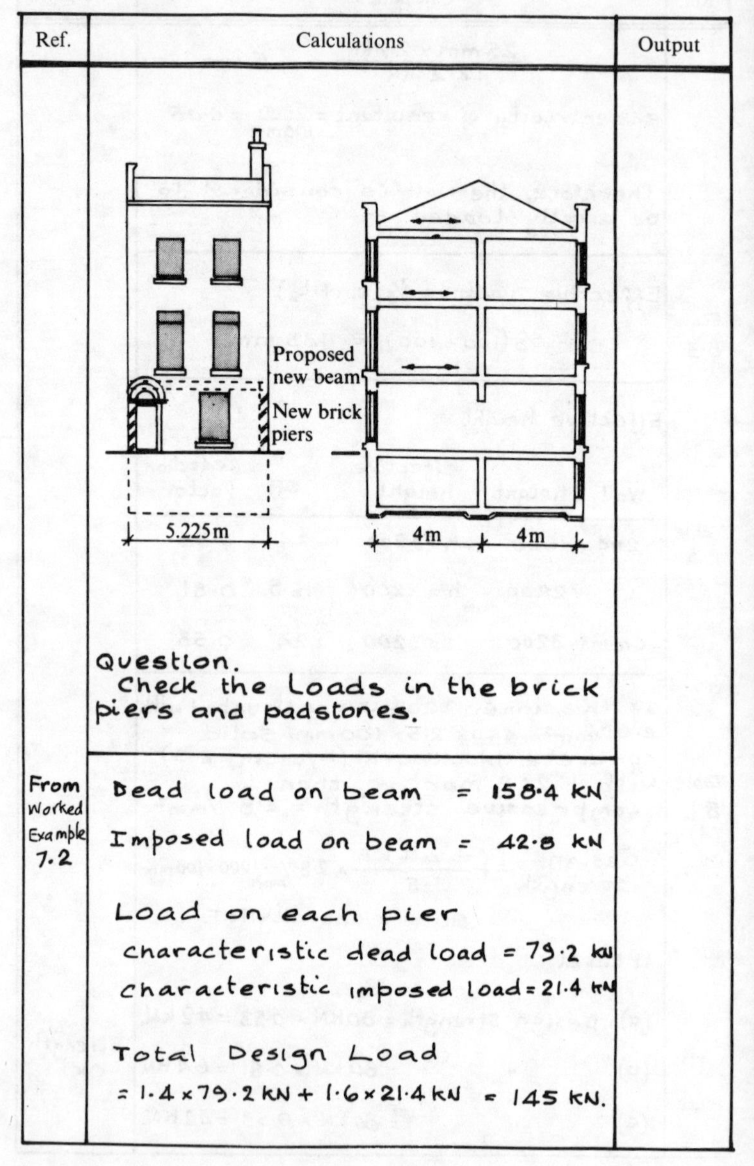

Ref.	Calculations	Output
	Question. Check the loads in the brick piers and padstones.	
From Worked Example 7.2	Dead load on beam = 158·4 kN Imposed load on beam = 42·8 kN Load on each pier characteristic dead load = 79·2 kN characteristic imposed load = 21·4 kN Total Design Load = 1·4 × 79·2 kN + 1·6 × 21·4 kN = 145 kN.	

Worked Example 8.2 continued

Ref.	Calculations	Output
Table 8.3	Capacity Reduction Factor = 1·00 The new brick pier will be laced into the existing brickwork of the party wall which will produce a stiff bruttress.	
clause 23.1.1 Ref. 8.1.	Area of Pier Assume area of pier 228 × 342 mm $A = 0{\cdot}08\,m^2$ which is less than $0{\cdot}2\,m^2$ ∴ Reduction in strength = $(0{\cdot}7 + 1{\cdot}5A) = 0{\cdot}82$	
Table 8.1 (a)	NEW BRICK PIERS Try class 5 Eng. bricks $35\,N/mm^2$ with 1:1:6 mortar. Characteristic compressive strength $= 0{\cdot}82 \times 8{\cdot}5\,N/mm^2 = 7\,N/mm^2$ Design strength of pier $= \frac{\text{Capacity R.F.}}{3{\cdot}5} \times 7\,N/mm^2 \times A$ $= \frac{1{\cdot}00}{3{\cdot}5} \times 7\,N/mm^2 \times 228 \times 342\,mm$ = 156 kN. which is greater than the total design load.	Strength OK.
	PADSTONES Concrete padstones are required on top of the piers. to distribute the load from the beam to the brickwork. Concrete padstones 228 × 342 × 90 mm Grade C30 concrete.	

CHAPTER 9

REINFORCED CONCRETE BEAM DESIGN

In the chapters covering timber and steel, the elastic method of structural design has been used. In this chapter and the one on masonry, the limit state method is used. The two methods were discussed in section 1.3, and it is advisable to refer to that section and also to section 8.5 on the structural design of masonry.

9.1 GENERAL REQUIREMENTS OF LIMIT STATE DESIGN

Limit state design considers all aspects of ultimate safety and functional requirements. Each aspect is called a limit state, so when designing a concrete beam, all the possible limit states should be considered. The structure must be safe, it must be serviceable in that it must function properly with no excessive deflections and no cracking or defects, which will impair the visual finish. It must, of course, be stable, durable and meet the required fire resistance. Each limit state has to be dealt with separately, and each one will have to meet the minimum requirements. If one limit state is not met, then the whole design will fail.

The various limit states are:

1. Strength
   (*a*) of each structural element
   (*b*) of the various materials used
2. Stability
   (*a*) of overall building structure
   (*b*) of individual elements
3. Serviceability (so a building can function during its lifetime)
   (*a*) deflection
   (*b*) cracking
   (*c*) dynamic effects
   (*d*) durability
   (*e*) fire resistance

9.2 LOADS AND SAFETY FACTORS

Design load = Safety factor × Characteristic load

The characteristic load is the actual load or the estimated likely load to be resisted by the structure. The safety factor will depend upon the combination of loads and how they may affect the structure. This is well illustrated when dead and wind load only are being considered. For example, the wind load on a roof may well be upwards, causing the roof to lift, so that the amount of dead load helping to hold the roof down may be a critical factor. To ensure that the roof stays in place, this dead load must not be overestimated, so in this situation, the recommended safety factor for the dead load is 1.0, and for the wind load, 1.4.

The safety factors recommended in reference 9.1 (at the end of the chapter) are:

(*a*) For strength limits

(i) Dead and imposed load only

Design dead load = 1.4 × Characteristic dead load
Design imposed load = 1.6 × Characteristic imposed load

(ii) Dead and wind load only

Design dead load = 1.0 × Characteristic dead load
Design wind load = 1.4 × Characteristic wind load

The values of typical characteristic dead and imposed loads are to be found in Appendices 2 and 3. Characteristic wind loads can be found from reference 9.3.

(*b*) For serviceability limits

(i) Dead and imposed loads only

Design dead load = 1.0 × Characteristic dead load
Design imposed load = 1.0 × Characteristic imposed load

(ii) Dead and wind load only

Design dead load = 1.0 × Characteristic dead load
Design wind load = 1.0 × Characteristic wind load

9.3 STRENGTH OF MATERIALS

Reinforced concrete consists of two materials, concrete and steel. The safety factors used when assessing the strength of these materials are 1.5 for concrete and 1.15 for steel. In practice it is not necessary to remember these values as they are already incorporated into the design charts which follow.

The characteristic strength of concrete, otherwise known as the cube strength, is the strength of a cube of concrete tested to destruction when it is 28 days old.

A selection of three grades of concrete are given in Table 9.1,

TABLE 9.1 Characteristic strength of concrete

Grade of concrete	*Cube strength (N/mm^2)*	*Characteristic strength (N/mm^2)*
C30	30	30
C35	35	35
C40	40	40

together with their characteristic strengths. The most appropriate grade of structural concrete for a small contractor to use is grade 25. Therefore, the design charts included in this chapter are for this grade of concrete only. If other grades of concrete are to be used, then reference 9.2 has to be consulted.

TABLE 9.2. Characteristic strength of reinforcement

Type of steel	*Characteristic steel strength (N/mm^2)*
Hot rolled mild steel	250
High yield steel (hot rolled or cold worked)	460

After Table 3.1, BS 8110 Part 1.

The characteristic strengths of steel reinforcement (see Table 9.1) is as specified in the appropriate British Standard. These are based on the ultimate strength of the steel. The worked examples and design charts in this Chapter have been restricted to the use of mild steel. If high-strength steels are to be used, then reference 9.1 should be consulted.

9.4 PROTECTION OF STEEL REINFORCEMENT

When steel is used in a building, it has to be protected against corrosion and fire. With reinforced concrete members, the steel reinforcement is protected by the concrete cover, the thickness of

which will depend upon the durability required against exposure to the weather and the resistance required against fire.

Protection against exposure

Table 9.3 gives the nominal cover to a selection of grades of concrete and conditions of exposures. This table has been based on Table 3.4 in reference 9.1.

TABLE 9.3. Durability (nominal cover to all reinforcement including the links)

Condition of exposure	*Nominal concrete cover (mm)* Grade 30	Grade 35	Grade 40
Mild			
Protected against the weather	25	20	20
Moderate			
Buried concrete	—	35	30
Sheltered from severe rain			
Severe			
Exposed to driving rain	—	—	40
Maximum free water/ cement ratio	0.65	0.60	0.55
Minimum cement ratio	275	300	325 kg/m^3

After Table 3.4, BS 8110 Part 1.

TABLE 9.4. Fire resistance

Description	*Concrete cover to give fire resistance in hours (mm)* $1\frac{1}{2}$ *hr*	*1 hr*	$\frac{1}{2}$ *hr*
Exposed concrete (siliceous aggregate)	25	20	20

After Table 3.5, BS 8110 Part 1.

Protection against fire

Table 9.4 is a simplified approach to the problem of fire, as there are a number of factors which could influence the fire resistance of a concrete structure in a building. Basically, concrete made with siliceous aggregates tend to spall at high temperatures and if this spalling can be retarded, the fire resistance of the element will increase. For example, if a concrete element is encased with plaster or vermiculite, then there will be an increase in its fire resistance.

9.5 DESIGN PARAMETERS FROM BS 8110 (REFERENCE 9.1)

1. Effective span. Clause 3.3.1.1
2. Minimum reinforcement. Clause 3.11.4.2
3. Maximum reinforcement. Clause 3.11.5
4. Minimum area of links. Clause 3.11.4.3

The reference clauses are the clauses from CP 110.

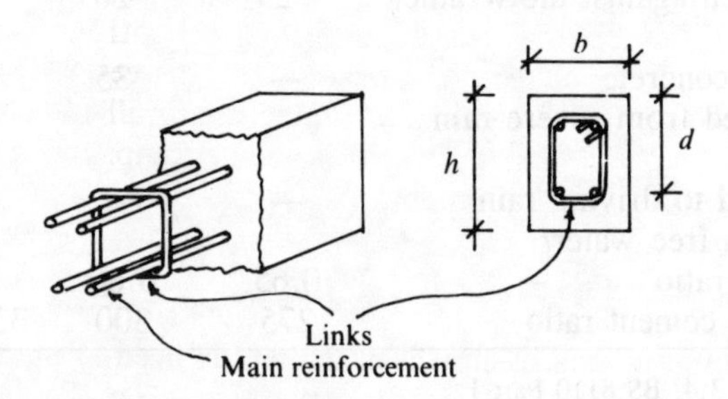

FIG. 9.1. Reinforcement in a beam

Effective span

The effective span will depend upon how the beam is supported and the amount of bearing. For a simply supported beam, the effective span is the smaller of:

The distance between the centre of bearings

OR

The clear distance between supports plus the effective depth of the beam

In practice this means that where the supports are very narrow, the beam will rotate about the centre of the supports, but where the supports are very wide, the ends of the beam will tend to lift when the beam begins to deflect and thus reduce the bearing on the support. Under these conditions, the centre of bearing is no longer at the centre of the support, so the effective span will be the second option.

Minimum reinforcement

The minimum steel reinforcement in a beam or slab where mild steel reinforcement is used, will be based on a percentage of the cross-sectional area of the beam (see Fig. 9.1, rectangular shape).

Minimum area of main reinforcement for mild steel $=0.24\%\,bh$

where h = overall depth of section
b = breadth of section

Minimum area of secondary reinforcement for mild steel in a solid concrete suspended slab should also be $0.24\%\,bh$.

The minimum reinforcement is a practical requirement allowing the concrete beam to withstand small impacts and generally holding the concrete together.

Maximum reinforcement

The maximum steel reinforcement is based on the practicalities of trying to pour concrete into formwork filled with steel bars. The aggregates must be able to flow around and between the bars, and it must be possible to compact the concrete into all the corners. To achieve this, the maximum area of tension or compression reinforcement must not exceed 4 per cent of the gross cross-sectional area of the beam.

Link reinforcement

The minimum area of links is based on the need to hold the main reinforcement in position and generally hold the concrete together. Along with the main reinforcement, the links provide a cage which gives the beam some resilience against accidental damage. Where the shear stress is significant, the links are there to resist the shear, but where the shear is not significant, nominal links are required.

Minimum size of link bars = at least ¼ size of main reinforcement
Maximum spacing = 12 times smallest compression bar

OR

= 0.75 times effective depth of beam

Nominal links for mild steel

$$\frac{A_{sv}}{S_v}=0.002\,b_v$$

where A_{sv} = the cross-sectional area of the two legs of the link

b_v = the breadth of the beam

S_v = the spacing of the links

9.6 SHEAR STRESS

The shear force in a beam can be found from the shear force diagrams as described in section 3.3. The shear stress at any cross-section will be:

$$\text{Shear stress} = \frac{\text{Shear force}}{\text{Effective area of section}}$$

For a rectangular concrete beam, the effective cross-sectional area is the breadth of the section times the effective depth, and the critical shear stress is the maximum value due to the maximum shear force at ultimate load.

$$\text{Max. shear stress } (v) = \frac{\text{Max. shear force } (V)}{\text{Breadth } (b) \times \text{depth } (d)}$$

If v exceeds v_c from Table 9.5, then shear reinforcement in the form of links is required, and if less, then nominal links should be used. The maximum shear stress must never exceed 3.75 N/mm² for grade 25 concrete.

For mild steel links, the relationship between the cross-sectional area of the two legs of a link (A_{sv}) and the spacing of the links (S_v) is found from the following equation:

$$\frac{A_{sv}}{S_v} \geq \frac{b_v(v - v_c)}{218}$$

where the measured units are: A_{sv} (mm²); S_v (mm); b_v (mm); v (N/mm²); v_c N/mm².

TABLE 9.5. Design concrete shear stress

Percentage of tension reinforcement at section $\left(\frac{100 A_s}{b_v d}\right)$	*Effective depth (mm)*		
	125	175	225
	v_c	v_c	v_c
(%)	*(N/mm²)*	*(N/mm²)*	*(N/mm²)*
0.25	0.53	0.49	0.46
0.50	0.67	0.62	0.58
1.00	0.84	0.78	0.73
2.00	1.06	0.98	0.92

After Table 3.9, BS 8110 Part 1.

It should be remembered that the spacing of the links (S_v) should not exceed 0.75d.

9.7 DESIGN CHARTS

The area of tension and compression reinforcement required in a reinforced concrete beam are found from design charts, similar to the one shown in Fig. 9.2. There is a different design chart for every combination of concrete grade and type of reinforcement, enough in fact to fill a book produced as a British Standard (reference 9.2 (BS 8110: Part 3)). Two charts have been selected from this publication,

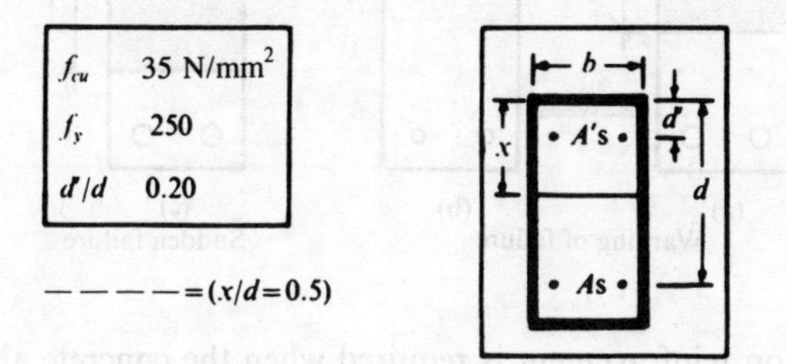

Design load = 1.4 characteristic dead load
+ 1.6 characteristic imposed load

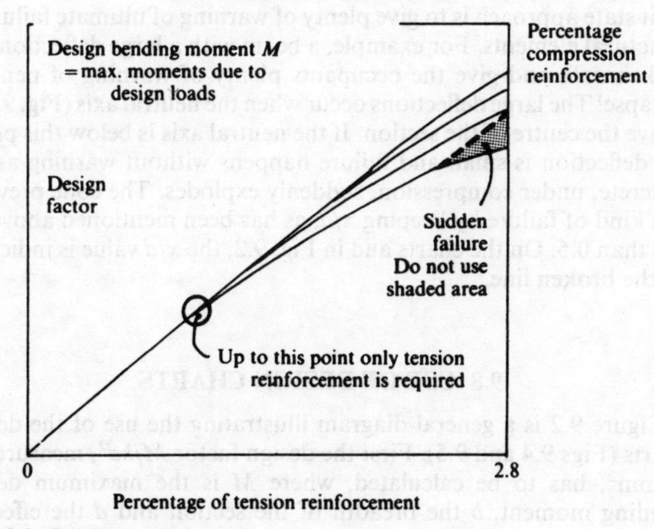

Percentage of tension reinforcement

FIG. 9.2. Guide to design chart

Figs. 8.4 and 8.5, which are for grade 35 concrete and high yield reinforcement. They cannot be used for any other grade of concrete or quality of steel reinforcement.

There is no difference in the two charts over the first part of the curve where tensile reinforcement only is required. Once compression reinforcement has to be added, the (d'/d) ratio begins to affect the graph, and the difference between the two charts becomes apparent. The d' is the depth to the compression reinforcement and d is the depth to the tensile reinforcement.

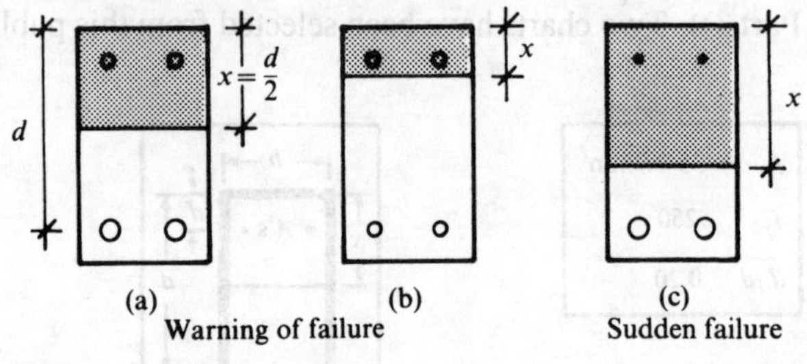

FIG. 9.3

Compression reinforcement is required when the concrete above the neutral axis is not strong enough to carry all the compression. In this respect, the distance to the neutral axis must not be greater than half the depth to the tensile reinforcement. This requirement is based on the philosophy of safety as discussed in section 9.1. The aim of the limit state approach is to give plenty of warning of ultimate failure of structural elements. For example, a beam with a large deflection will look unsafe and give the occupants plenty of warning of pending collapse! The large deflections occur when the neutral axis (Fig. 9.3) is above the centre of the section. If the neutral axis is below this point, the deflection is small and failure happens without warning as the concrete, under compression, suddenly explodes. The code prevents this kind of failure by keeping x/d, as has been mentioned above, to less than 0.5. On the charts and in Fig. 9.2, the x/d value is indicated by the broken line.

9.8 USE OF DESIGN CHARTS

Figure 9.2 is a general diagram illustrating the use of the design charts (Figs 9.4 and 9.5). First the design factor M/bd^2, measured in N/mm^2, has to be calculated, where M is the maximum design bending moment, b the breadth of the section and d the effective depth. As b and d are initially unknown, an intelligent guess has to be

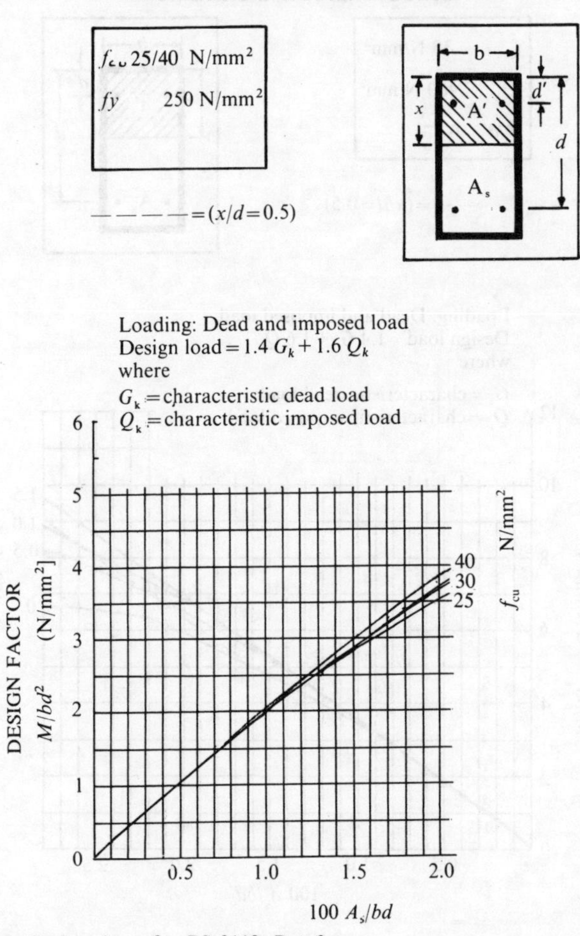

FIG. 9.4. Design chart after BS 8110: Part 3

made as regards to their values by using the rules of thumb given in section 1.4. The percentage of tensile reinforcement required ($100A_s/bd$) is then read from the graph. For example, if the design factor equals 1.5, then from the charts the percentage of tensile reinforcement equals 0.75. The area of the reinforcement will, therefore, be 0.75 per cent of the effective cross-sectional area of the beam.

If the design factor comes out to the higher, say 3, then there will be more than one line on the chart to choose from. The different lines indicate different amounts of compression reinforcement, so a decision has to be made as to the amount of compression reinforce-

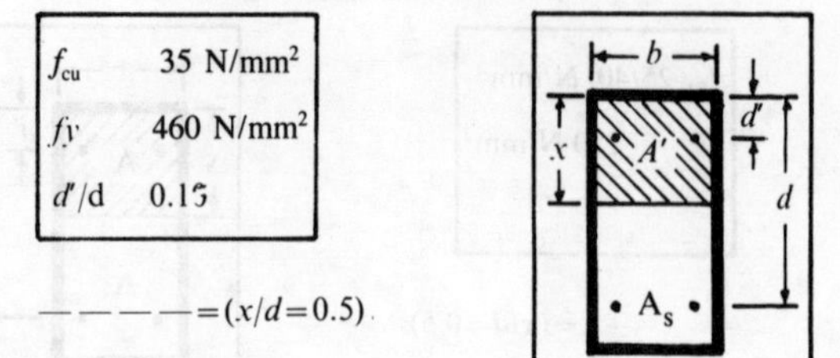

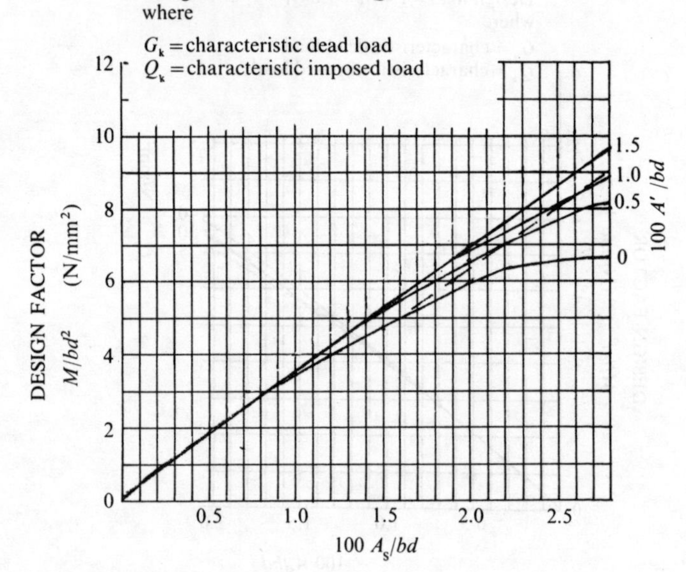

FIG. 9.5. Design chart after BS 8110: Part 3

ment to be used, before the percentage of tensile reinforcement required can be read from the chart. The percentage of compression reinforcement will take into account the practicalities of placing at least two longitudinal bars in the top of the beam, and the percentage of this reinforcement will depend upon the cross-sectional area of the bars.

If the initial guess as to the dimensions of the beam are wrong, then it will be obvious from the chart. The design factor will be far too large or far too small, and another guess will have to be made and the procedure repeated.

Once the percentages of reinforcement have been found, the

required cross-sectional area of reinforcement has to be calculated, and the bar sizes read from Table 9.6. This table tabulates the cross-sectional area for various combinations of number of bars and sizes of bars. For example, if the area of reinforcement required is 1500 mm², then the table shows that 5 number 20 mm diameter bars gives an area 1570 mm², or 2 number 32 mm diameter bars give an area of 1610 mm². The designer has to make the choice which one to use.

Table 9.7 has been prepared for reinforced concrete slabs, where the reinforcement required is based on a metre width of slab. The designer here has to make the decision between the sizes of bars and their spacing. For example, if the area required per metre width is 500 mm², then:

8 mm dia. bars at 200 mm centres gives an area of 503 mm²
10 mm dia. bars at 150 mm centres gives an area of 524 mm²
12 mm dia. bars at 200 mm centres gives an area of 565 mm²

Table 9.8 gives a selection of steel mesh sizes which can be used instead of Table 9.7. Many small builders prefer to use mesh reinforcement when constructing reinforced concrete slabs. Thus, if the area required per metre width is 500 mm², then:

B 503 structural mesh fabric gives an area of 503 mm²

REFERENCES

9.1 *Structural Use of Concrete,* British Standard 8110: Part 1, 1985, Design and Construction.
9.2 *Structural Use of Concrete,* British Standard 8110: Part 3, 1985, Design Charts.
9.3 *Design Loading for Buildings,* British Standard 6399: Part 2, Code of Practice for Wind Loads.
9.4 *Steel Fabric for the Reinforcement of Concrete,* British Standard 4483.

TABLE 9.6. Areas of steel reinforcement (for beams)

Number of bars	*Cross-sectional areas for specified number of bars (mm^2)*							
	6 mm	*8 mm*	*10 mm*	*12 mm*	*16 mm*	*20 mm*	*25 mm*	*32 mm*
1	28.3	50.3	78.5	113	201	314	491	804
2	56	100	157	226	402	628	981	1608
3	84	150	235	339	603	942	1472	2412
4	113	201	314	452	804	1256	1963	3216
5	141	251	392	565	1005	1571	2454	4021
6	169	301	471	678	1206	1885	2945	4825
7	198	352	549	791	1407	2199	3436	5629
8	226	402	628	904	1603	2513	3927	6433
9	254	452	706	1017	1809	2827	4418	7237
10	283	503	785	1131	2011	3142	4909	8042

TABLE 9.7. Areas of steel reinforcement (for slabs)

Diameter of bars	*Cross-sectional areas per metre width at various spacings (mm^2)*						
	Spacing of bars						
	75 mm	*100 mm*	*125 mm*	*150 mm*	*175 mm*	*200 mm*	*300 mm*
6 mm	377	283	226	188	161	141	94
8 mm	670	503	402	335	287	251	167
10 mm	1046	785	628	523	448	392	261
12 mm	1508	1131	904	754	646	565	377
16 mm	2681	2011	1608	1340	1149	1005	670
20 mm	4189	3142	2513	2094	1795	1571	1047

TABLE 9.8. A selection of steel mesh sizes

BS reference	Mesh sizes (nominal pitch of wires) (mm)		Wire sizes (mm)		Cross-sectional area per metre width (mm^2)		Nominal mass per square metre (kg)
	Main	Cross	Main	Cross	Main	Cross	
			Square mesh fabric				
A 393	200	200	10	10	393	393	6.16
A 252	200	200	8	8	252	252	3.95
A 193	200	200	7	7	193	193	3.02
A 142	200	200	6	6	142	142	2.22
A 98	200	200	5	5	98.0	98.0	1.54
			Structural mesh fabric				
B 1131	100	200	12	8	1131	252	10.9
B 785	100	200	10	8	785	252	8.14
B 503	100	200	8	8	503	252	5.93
B 385	100	200	7	7	385	193	4.53
B 283	100	200	6	7	283	193	3.73
B 196	100	200	5	7	196	193	3.05

Sizes of sheets 4.8 m × 2.4 m.
After Tables 1 and 3, BS 4483.

Worked Example 9.1
Project: Detached house
Suspended floor slab

Ref.	Calculations	Output
	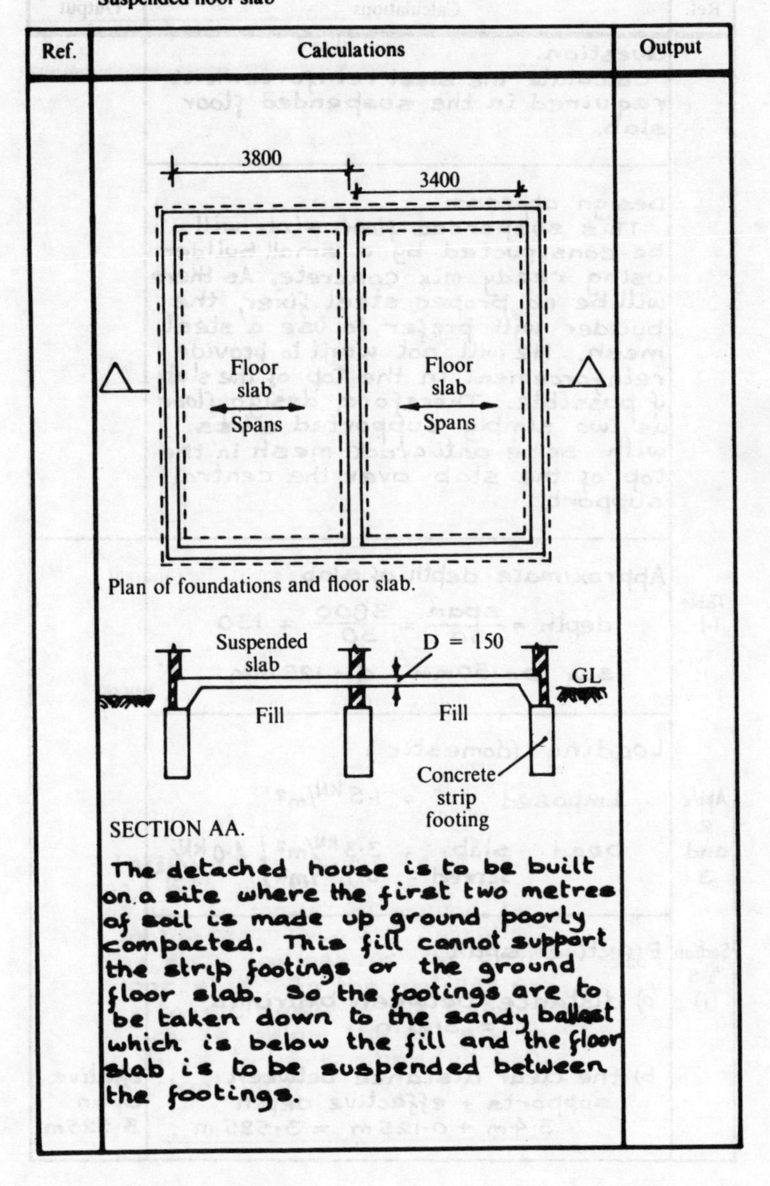 Plan of foundations and floor slab. SECTION AA. The detached house is to be built on a site where the first two metres of soil is made up ground poorly compacted. This fill cannot support the strip footings or the ground floor slab. So the footings are to be taken down to the sandy ballast which is below the fill and the floor slab is to be suspended between the footings.	

Worked Example 9.1 continued

Ref.	Calculations	Output
	Question. Calculate the steel reinforcement required in the suspended floor slab.	
	Design approach This suspended floor slab will be constructed by a small builder using ready mix concrete. As there will be no proper steel fixer, the builder will prefer to use a steel mesh. He will not want to provide reinforcement in the top of the slab, if possible. Therefore, design floor as two simply supported slabs with some anti-crack mesh in the top of the slab over the central support.	
Table 1.1	**Approximate depth of slab** $\text{depth} = \frac{\text{span}}{30} = \frac{3800}{30} = 130$ say $D = 150$ mm $d = 125$ mm	
App'x 2 and 3	**Loading (domestic)** Imposed = $1.5\ \text{kN/m}^2$ Dead slab = $3.3\ \text{kN/m}^2$, screed = $0.7\ \text{kN/m}^2$ } $4.0\ \text{kN/m}^2$	
Section 9.5 (i)	**Effective span** a) distance between bearings = 3.8 m b) the clear distance between supports + effective depth 3.4 m + 0.125 m = 3.525 m	Effective span 3.525 m

Worked Example 9.1 continued

Ref.	Calculations	Output
Table 9.3	Concrete cover (C35 GRADE) Assume that the underside of the slab is protected from the ground by a layer of 'blinding' concrete, and therefore, concrete cover required = 20mm.	
	Design Loads Consider a 1m wide strip of slab spanning 3.525 m. Total characteristic imposed load $= 1.5 kN/m^2 \times 1\,m \times 3.525\,m = 5.3\,kN$ Total characteristic dead load $= 4.0 kN/m^2 \times 1\,m \times 3.525\,m = 14.1\,kN$ Design load $= 1.4 \times 14.1\,kN + 1.6 \times 5.3\,kN$ $= 19.7 + 8.5 = 28.3\,kN$	Design Load 28.3 kN
Worked Example 3.3 Fig. 9.4 Table 9.7 Table 9.8	Max. BM $M = \frac{WL}{8} = \frac{28.2\,kN \times 3.525\,m}{8}$ $= 12.4\ kN.m$ Design factor $\frac{M}{bd^2} = \frac{12\,400\,000}{1000 \times 125 \times 125} = 0.8$ $100\frac{A_s}{bd} = 0.4$ (from graph) Tensile steel $= A_s = \frac{0.4}{100} \times 1000 \times 125$ $= 500\ mm^2$ 10mm dia. bars at 150mm centres OR Structural mesh B503 (main 503mm² cross 252mm²)	Mesh B503

Worked Example 9.1 continued

Ref.	Calculations	Output
Section 9.5 (ii)	Minimum reinforcement Min. main reinforcement = 0·24% effective cross-sectional area $= \frac{0.24}{100} \times 1000\ mm \times 150\ mm = 360\ mm^2$ (Less than moment requirement – ignore) Min. secondary reinforcement (suspended concrete slab) = 0·24% gross cross-sectional area $= \frac{0.24}{100} \times 1000\ mm \times 150\ mm = 360\ mm^2$ Steel mesh A 393 ok (cross 393 mm^2)	
BS8110 clause 3·4·6 BS8110 Table 3·10 BS8110 Table 3·11	Check deflection. Deflection is not usually a problem when mild steel reinforcement and grade 35 concrete is used. Deflection is limited to $\frac{span}{250}$ $\frac{\text{basic span}}{\text{effective depth}}$ × Modification factor Ratio = 20 for simply supported $\frac{100 A_s}{bd} = 0.4$ Service stress = 145 N/mm^2 (mild steel) Mod. factor = 2·00	
	Span/effective ratios Permissible = 2·00 × 20 = 40 Actual $= \frac{3.525}{0.125} = 28.6$ OK.	Deflection OK.

Worked Example 9.2
Project: Terrace house
Calculations for concrete beam

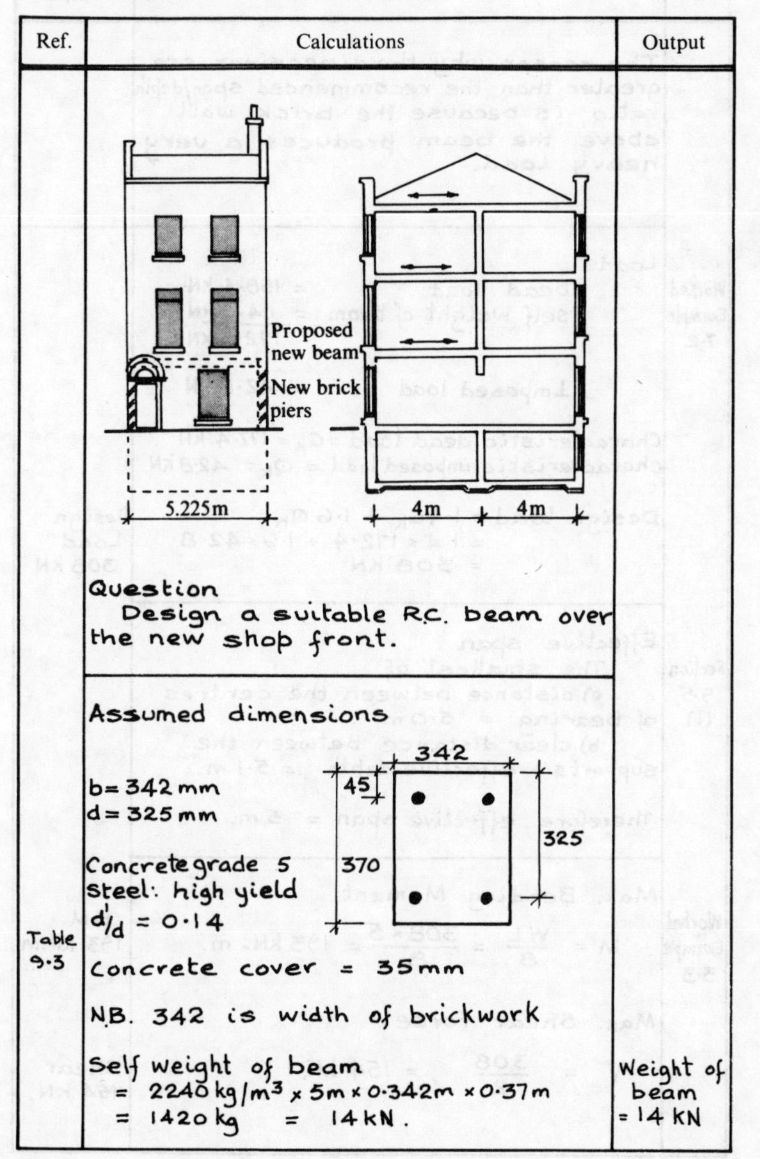

Ref.	Calculations	Output
	Proposed new beam New brick piers 5.225m 4m 4m Question Design a suitable R.C. beam over the new shop front.	
Table 9.3	Assumed dimensions b = 342 mm d = 325 mm Concrete grade 5 Steel: high yield $d'/d = 0.14$ 342 45 325 370 Concrete cover = 35mm N.B. 342 is width of brickwork Self weight of beam = 2240 kg/m^3 × 5m × 0·342m × 0·37m = 1420 kg = 14 kN.	Weight of beam = 14 kN

Worked Example 9.2 continued

Ref.	Calculations	Output
	The reason why the dimensions are greater than the recommended span/depth ratio, is because the brick wall above the beam produces a very heavy load.	
Worked Example 7.2	Loads Dead load = 158.4 kN self weight of beam = 14.0 kN 172.4 kN Imposed load = 42.8 kN Characteristic dead load = G_K = 172.4 kN Characteristic imposed load = Q_K = 42.8 kN Design Load = $1.4G_K + 1.6Q_K$ $= 1.4 \times 172.4 + 1.6 \times 42.8$ = 308 KN	Design Load 308 kN
Section 9.5 (i)	Effective span The smallest of a) distance between the centres of bearing = 5.0 m b) clear distance between the supports + effective depth = 5.1 m. Therefore effective span = 5 m.	
Worked Example 3.3	Max. Bending Moment $M = \frac{WL}{8} = \frac{308 \times 5}{8} = 193$ kN. m. Max. Shear Force $V = \frac{308}{2} = 154$ kN.	BM 193 kN.m Shear 154 kN.

Worked Example 9.2 continued

Ref.	Calculations	Output
Fig 9.5	Design factor $d'/d \doteq 0.15$ $\frac{M}{bd^2} = \frac{193\ 000\ 000 \text{ N.mm}}{342 \times 325 \times 325} = 5.3$ $100\frac{A_s}{bd} = 1.55 \quad 100\frac{A_s'}{bd} = 0.5$	
Table 9.6	Tension reinforcement $A_s = \frac{1.55}{100} \times bd = \frac{1.55}{100} \times 342 \times 325$ $= 1730 \text{ mm}^2$ 4 no. 25 mm diameter bars	4 no T25 (1963 mm^2)
Table 9.6	Compression reinforcement $A_s' = \frac{0.5}{100} \times bd = \frac{0.5}{100} \times 342 \times 325$ $= 560 \text{ mm}^2$ 2 no. 25 mm diameter bars	2 no. T 25 (981 mm^2)
Section 9.6 Table 9.5	Shear stress v $v = \frac{\text{Shear Force}}{\text{effective area of section}}$ $= \frac{154 \text{ kN}}{bd} = \frac{154\ 000}{342 \times 325} = 1.38 \text{ N/mm}^2$ Assume 2 no. 25 mm bars extend to end of beam. $A_s = 981 \text{ mm}^2$ $100\frac{A_s}{bd} = \frac{100 \times 981}{342 \times 325} = 0.88$ $\therefore v_c = 0.73 \text{ N/mm}^2$ $v - v_c = 1.38 - 0.73 = 0.65$	$v = 1.38 \text{ N/mm}^2$ $v_c = 0.73 \text{ N/mm}^2$

Worked Example 9.2 continued

Ref.	Calculations	Output
Section 9.6	Links $\frac{A_{sv}}{S_v} = \frac{b(v - v_c)}{218} = \frac{342 \times 0.65}{218} = 0.99$ A_{sv} = area of two legs of a link S_v = spacing of links If the spacing S_v = 200 mm $A_{sv} = 200 \times 0.99 = 198\ mm^2$ Say 12 mm links at 200 mm centres	12 mm Links at 200 mm centres
Section 9.5 (iv)	Nominal links $\frac{A_{sv}}{S_v} = 0.002\ b_v$ Max. spacing of links $= 0.75 \times 325 = 240$ mm $A_{sv} = 0.002 \times 342 \times 240 = 164\ mm^2$ Say 12 mm links at 240 mm centres	12 mm Links at 240 mm Centres
	Distribution of shear stress along beam. 1.38 0.75 + − x 2.5m 1.38 $\frac{x}{0.75} = \frac{2.5\ m}{1.38}$ $x = 1.36$ Distribution of 12mm dia. links. 1300 2640 1300 Links 12mm dia. at 200mm Centres Links 12mm dia. at 240 mm centres	

Worked Example 9.2 continued

Ref.	Calculations	Output
	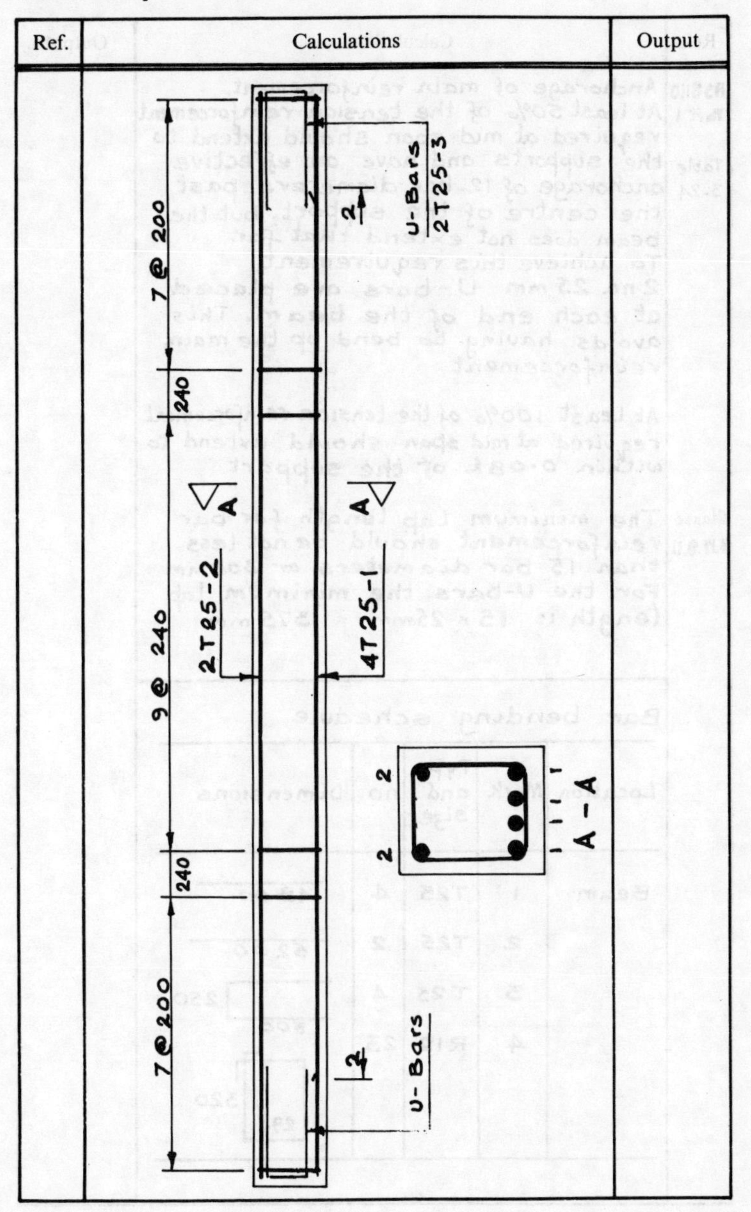	

Worked Example 9.2 continued

Ref.	Calculations	Output
BS8110 Part 1 Table 3.24	Anchorage of main reinforcement. At least 50% of the tension reinforcement required at mid span should extend to the supports and have an effective anchorage of 12 bar diameters past the centre of the support, but the beam does not extend that far. To achieve this requirement, 2 no. 25 mm U-bars are placed at each end of the beam. This avoids having to bend up the main reinforcement. At Least 100% of the tension reinforcement required at mid span should extend to within 0·08L of the support.	
Clause 3.12.8.11	The minimum lap length for bar reinforcement should be not less than 15 bar diameters or 300 mm For the U-bars the minimum lap length is 15 × 25mm = 375 mm	

Bar bending schedule

Location	Mark	Type and size	no	Dimensions
Beam	1	T25	4	4800
	2	T25	2	5200
	3	T25	4	500, 250
	4	R12	23	290, 320

Appendix 1

SI UNITS

The name 'Système International d'Unités' (International System of Units), with the abbreviation SI, was adopted by the Eleventh General Conference on Weights and Measures in 1960.

Table A1.1. Base Units

Quantity	*Name of unit and symbol*
Length	Metre (m)
Mass	Kilogram (kg)
Time	Second (s)

Table A1.2. Derived units

Physical quantity	*Unit and symbol*	*Derivation*	*Definition*
Force	Newton (N)	$kg \cdot m/s^2$	The newton is the force which when applied to a body having a mass of 1 kilogram, causes an acceleration of 1 metre per second per second in the direction of application of the force. (With the earth's gravitational pull being $9.81\,m/s^2$ 1 kg will produce a force of 9.81 N.)

TABLE A1.3. A selection of SI units and their decimal multiples

Physical quantity	*Unit*	*Decimal multiples*	
Length	m	km	Kilometre
		m	Metre 1000 mm = 1 m
		mm	Millimetre
Area	m^2	m^2	Square metre 1 000 000 mm^2 = 1 m^2
		mm^2	Square millimetre
Volume	m^3	m^3	Cubic metre 1 000 000 000 mm^3 = 1 mm^3
		mm^3	Cubic millimetre
Mass	kg	Mg	Megagram 1000 kg = 1 Mg
		kg	Kilogram
Acceleration	m/s^2	m/s^2	Metres per second per second
Density	kg/m^3	kg/m^3	Kilogram per cubic metre
Force	N	MN	Meganewton 1000 kN = 1 MN
		kN	Kilonewton 1000 N = 1 kN
		N	Newton 10 N approximately equals 1 kg (on earth 9.81 N = 1 kg)
Stress	N/mm^2	kN/mm^2	Kilonewton per millimetre square
		N/mm^2	Newton per millimetre square
		kN/m^2	Kilonewton per metre square
		N/m^2	Newton per metre square 1000 N/mm^2 = 1 kN/mm^2
	Pa	Pa	Pascal (used in New Zealand and Australia)
		MPa	Megapascal 1 MPa = 1 N/mm^2
Strain	No units		Unitless

APPENDIX 2

IMPOSED LIVE LOADS

TABLE A2.1. Imposed floor loads

Building use	*Intensity of distributed load (kN/m^2)*
Classrooms	3.0
Bedrooms	1.5
Gymnasia	5.0
Houses	1.5
Office: general	2.5
filing and storage	5.0
Libraries: reading rooms	2.5
stack rooms	6.5 minimum

TABLE A2.2. Imposed roof loads (snow loads)

Slope	*Intensity of distributed load (kN/m^2)*	
0 --→ 10°	0.72	For less than three storeys and single occupation
	0.75	No access
	1.5	With access
10 --→ 30°	0.75	
30 --→ 75	0.75 --→ 0	

TABLE A2.3. Parapets and balustrades

Element	*Horizontal force (kN/m run)*
Landings and balconies – domestic	0.36
– others	0.74
Panic barriers	3.0

TABLE A2.4. Imposed ceiling loads

Ceiling loads	0.72 kN/m^2	Water tanks distributed over a number of joists

TABLE A2.5. Reductions in imposed loads for number of floors

Number of floors, including the roof, carried by member	*Reduction in imposed load (%)*
1	0
2	10
3	20
4	30

Wind loads — see British Standard 6399, Part 2.
References: Tables A2.1 to A2.5 are after British Standard 6399, Part 1: Dead and imposed loads. The loading 0.72 kN/m^2 in Table A2.2 is after the Building Regulations.

APPENDIX 3

DEAD LOADS

TABLE A3.2. Materials

Materials		*Distributed load*
Reinforced concrete	Cubic metre	24 kN/m^3
Concrete screed	10 mm thickness	0.26 kN/m^2
Clay floor tiles		0.6
Asphalt	10 mm thickness	0.25
Timber		
Douglas fir	Cubic metre	5.0 kN/m^3
Pitch pine	Cubic metre	7.0
Floorboards	21 mm thickness	0.11 kN/m^2
	29 mm	0.14
Chipboard	21 mm thickness	0.16
Plasterboard with	12 mm thickness	0.12
setting coat	18 mm	0.18
Brickwork	103 mm thickness	2.4
Windows		0.3–0.4
Slates and tiles		0.7–1.1

After BS 648, Schedule of weights of building materials.

TABLE A3.1. Building elements

Element	*Items included*	*Distributed load (kN/m^2)*	
Timber roofs	Tiles, battens, felt, trusses Plaster and skim ceiling	0.75–1.5 Measured in horizontal plane	
Timber floors	Floorboards, joists Plaster and skim ceiling	0.4–0.5	
Concrete floors	Solid slabs	75 mm deep	1.8
		100 mm	2.4
		150 mm	3.6
	Ribbed slabs	150 mm deep	2.5
		300 mm	3.5
Walls	225 mm brick (plaster and skim walls)	4.5–5.5	
	100 mm + 100 mm concrete blocks (plaster and skim walls)	Hollow	3.0
		Solid	4.5

APPENDIX 4

MAXIMUM BENDING MOMENTS, SHEAR FORCES AND DEFLECTIONS FOR SIMPLY SUPPORTED BEAMS

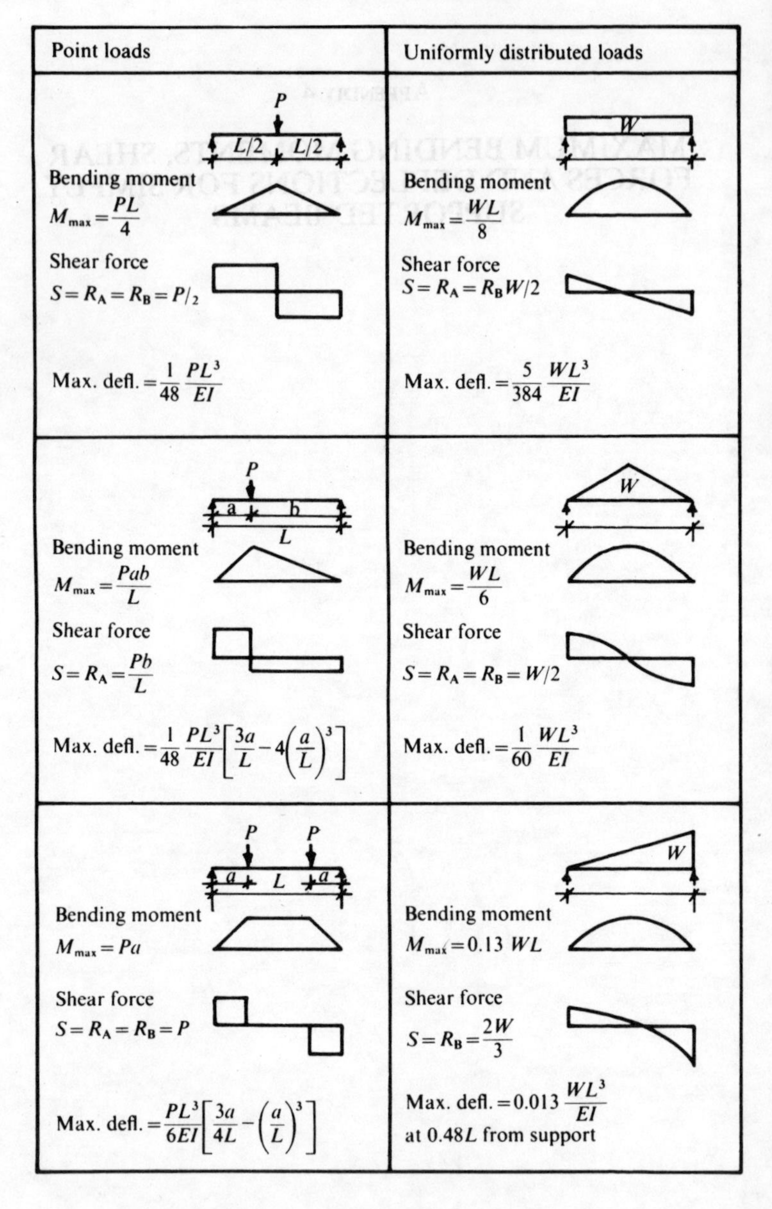

Point loads	Uniformly distributed loads
Bending moment $M_{max} = \frac{PL}{4}$ Shear force $S = R_A = R_B = P/2$ Max. defl. $= \frac{1}{48}\frac{PL^3}{EI}$	Bending moment $M_{max} = \frac{WL}{8}$ Shear force $S = R_A = R_B W/2$ Max. defl. $= \frac{5}{384}\frac{WL^3}{EI}$
Bending moment $M_{max} = \frac{Pab}{L}$ Shear force $S = R_A = \frac{Pb}{L}$ Max. defl. $= \frac{1}{48}\frac{PL^3}{EI}\left[\frac{3a}{L} - 4\left(\frac{a}{L}\right)^3\right]$	Bending moment $M_{max} = \frac{WL}{6}$ Shear force $S = R_A = R_B = W/2$ Max. defl. $= \frac{1}{60}\frac{WL^3}{EI}$
Bending moment $M_{max} = Pa$ Shear force $S = R_A = R_B = P$ Max. defl. $= \frac{PL^3}{6EI}\left[\frac{3a}{4L} - \left(\frac{a}{L}\right)^3\right]$	Bending moment $M_{max} = 0.13\,WL$ Shear force $S = R_B = \frac{2W}{3}$ Max. defl. $= 0.013\,\frac{WL^3}{EI}$ at $0.48L$ from support